Arthur Frederick Guy

Electric Light and Power

Giving the Result of Practical Experience in Central Station Work

Arthur Frederick Guy

Electric Light and Power
Giving the Result of Practical Experience in Central Station Work

ISBN/EAN: 9783337025205

Printed in Europe, USA, Canada, Australia, Japan

Cover: Foto ©berggeist007 / pixelio.de

More available books at **www.hansebooks.com**

ELECTRIC LIGHT AND POWER.

ELECTRIC LIGHT AND POWER

GIVING

THE RESULT OF PRACTICAL EXPERIENCE

IN

CENTRAL-STATION WORK.

BY

ARTHUR F. GUY, A.M.I.E.E.

ILLUSTRATED.

1894.

BIGGS AND CO., 139-140, SALISBURY COURT, FLEET STREET. E.C.

PREFACE.

MANY books are issued having electricity for their subject : some are mathematical, some school text-books, some theoretical, and some few practical. The accompanying volume will be given, the author hopes, a place among the last class, as throughout its pages he has striven to record and explain that information only which may be termed *useful practical knowledge.*

Where necessary, the principles underlying the action of apparatus have been discussed briefly, and only so far as to enable the reader to grasp the laws that govern their action, because a certain amount of theory is absolutely requisite to permit of their being used in an intelligent and efficient manner.

ARTHUR F. GUY.

ENGINEERING DEPARTMENT,
G.P.O., LONDON.

CONTENTS.

———◆———

ELECTRIC LIGHT AND POWER.

CHAPTER I.

EVOLUTION OF ELECTRICAL ENGINEERING.

The First Electric Light—Discovery of Dynamic Elec-
tricity—Evolution of the Modern Dynamo and Lamps—
Subsequent Progress — Electricity in England and
Abroad—Advantages of the Electric Light—Cost of
Electricity v. Cost of Gas—Economies of the Electric
Light—Sources of Power—Conservation of Energy.

The First Electric Light.

It was on one evening in the year 1810 that Sir Humphrey
Davy showed to a large and distinguished audience, assembled
at the rooms of the Royal Institution, the brilliant and
dazzling light that was produced by passing a current of
electricity through two pencils of charcoal, about an inch
long and one-sixteenth of an inch thick, the said pencils
being placed end to end, and slightly separated from each
other. This is generally conceded to have been the first time
that the world saw the electric light, hence we may date its
birth from the above historical event.*

---

* (*Extract from letter published in the "Electrical Engineer" of
September 16th, 1892, from* Mr. ARTHUR SHIPPEY.) " . . .
Mr. Guy fixes the adventual year of the evolution of electric
lighting as 1810. Now, I think I am right in suggesting that
this should be fixed as the year 1809, instead of 1810, especially

Discovery of Dynamic Electricity.

In those days the only available source of electricity was the "primary or voltaic cell," and to obtain any large quantity of electricity a great number of these had to be used. A number of cells connected together forms a "battery," probably so named after the example afforded by calling a number of field-guns a battery. The battery employed by Sir Humphrey Davy consisted of 2,000 cells, so some idea can be formed of the great trouble and expense incurred in handling such a quantity. A "voltaic cell," as most people know, consists of two different metals put into a glass or porcelain jar, the jar being divided by a porous partition, and filled with acidulated solutions. A wire is fixed to each metal plate, and when the two free ends of the two wires are joined together so as to form a complete circuit, then a current of electricity will flow, the strength of which will depend on the size and nature of the metals, and the kind of solution, etc.

It was reserved to the genius of those two seekers after truth, Henry and Faraday, to disclose to the world the great discovery that magnetism could produce electricity—a discovery that is justly classed as one of the most important that man ever made. The honour is equally divided between the two above-named men. Each was

as the first practical experiments relating to the production of electric light were carried out, not by Sir Humphrey Davy himself, but by Dr. Wollaston, Sir E. Horne, and Mr. Children, the designer of the Cruikshank-Children battery, as used for the experimental work carried out at the Royal Institution, in conjunction with Prof. Brands, the secretary of the above institution. . . . Amongst them [early experimentalists] I think should be coupled the name of Ritter, whose early and successful experiments on battery work and the storage of electricity, is perhaps one of the most interesting in these modern times. His early work is clearly described in Izarn's " Manuel du Galvanisme " (1805) and other French journals published between the years 1803 and 1806. . . . "

working patiently and laboriously to find the link between electricity and magnetism—Henry in the New World, amidst his busy work-a-day duties of teaching at Albany Academy, New York State, snatching, when he could; a few spare moments to spend on his absorbing experiments ; Faraday, in England, with ample leisure and scientific facilities of every description at his command, occupying, as he did, the position of Chemical Professor at the Royal Institution. Each was working independently of the other, and made the long-wished-for discovery independently. In point of time Henry was the first to reach the goal—August, 1830, being given as the time of his decisive experimental proof; while Faraday arrived at the same ends in September, 1831, ignorant of Henry's work.

And now to set forth the nature of this momentous discovery. Prior to this time it was known that when a current of electricity was flowing in a conductor or wire it would deflect the needle of a compass placed near to the wire, showing that the electric current acted magnetically by deflecting or attracting the needle. Henry and Faraday discovered the inverse of this—i.e., that magnetism could produce an electric current. When a piece of iron is magnetised its two ends are named the "poles"; when the iron bar is bent into the shape of a horseshoe its two ends are brought near together, so that both the poles can exert their force jointly. From one pole of the magnet a stream of what has been named by Faraday "lines of force" issue, and re-enter at the other pole; in the case of the horseshoe magnet, the lines curve across the air gap between the two poles, and the space which these lines of force occupy is named the "magnetic field." These lines of force, of course, are only imaginary, like the Equatorial line, but they serve to illustrate very

B 2

clearly the direction of the magnetic forces. When the
ends of the magnet do not face each other the lines curve
round in issuing and entering. If a length of wire or metal
conductor, whose two ends are joined so as to make a
complete circuit, be caused to move across the magnetic
field, so as to *vary* the number of the magnetic " lines of
force " that thread through or are embraced by the coil, then
a current of electricity will be generated in that conductor,
or coil, and will last so long as the movement of the conductor lasts in the magnetic field. The same effect will be
produced if the coil is stationary and the magnet be caused
to move.   This, then, in brief, is the fundamental principle
upon which all dynamos work, and without the possession
of this it is safe to say there would be no electric lighting as
it is to-day.

Evolution of the Modern Dynamo and Lamps.

For some considerable time the dynamo was confined to
the laboratory, and looked upon as a scientific piece of
apparatus, scientists little dreaming what a future it had
before it. The first practical form which the dynamo
assumed was that brought out by Dr. Werner Siemens in
1867, and during the subsequent 10 years many keen minds
were hard at work improving it, and gradually bringing it
into shape, electrically and mechanically.

Naturally, the next article that claims our attention is
the electric lamp, the adjunct to the dynamo, the evolution
of which was slow and gradual, like the dynamo. When a
current of electricity is passed through two pencils of charcoal
or carbon, slightly separated, light is emitted, and these
slowly consume away. The carbons used to be fed forward
by hand ; this primitive method was supplanted by automatic
feeding, effected by clockwork. The first arc lamp of real
practical value was designed in 1857 by Serrin, who caused

the electric current to regulate the lamp automatically. After this rapid improvements appeared, resulting in the excellent types of to-day.

Turning now to the glow or incandescent lamp, early experimentalists first used platinum wire sealed in a glass bulb, filled with non-combustible vapour. Upon sending a current through a thin piece of metal of high resistance, like platinum or iridium, it is raised to a white heat, the brilliancy or intensity of the light emitted depending on the quantity of current ; this is the principle of the incandescent lamp, and since the metal would consume away in the atmosphere it was necessary to place it in a non-combustible vapour, such as nitrogen. Through the invention of the air-pump, however, a new field presented itself. About 1878, Prof. W. Crookes made a series of valuable experiments on the vacuum tube, and conceived the idea of making incandescent lamps by placing the thin metal thread ·or filament in a glass bulb exhausted of air.

Simultaneously we had Edison working in America, and Lane, Fox, and Swan in England, each striving to evolve an incandescent lamp of commercial value ; each was successful, and at about the same time. Edison and Swan put the fruits of their labour together, and then was created the famous Edison-Swan lamp, consisting of a thin curled thread of carbon sealed in an air-exhausted glass bulb.

Subsequent Progress.

During the last few years invention has been going on at a very rapid rate. The most eminent scientists and physicists of the day have given their minds to the task, and have diligently laboured and experimented in the laboratory. Bit by bit, piece by piece, has the edifice of electrical science been built up. One man's mistake was another man's luck, defects were rooted out one by one, and no sooner was one thing accomplished than another man tried to improve it. The man of science was, however, not sufficiently backed up by practical aid. The early electricians did not exhibit good mechanical construction in their apparatus : they were ignorant of the forces with which they had to deal ; in one word, they did not treat electricity as *electric horse-power.*

Dynamos were built in too weak a manner ; there was no mechanical design about them ; a piece of string used to be put where a piece of steel wire is now put. It is now thoroughly understood that a dynamo, like any other machine which produces power, must à *priori* be designed and built in a strong and substantial mechanical fashion, just the same as a steam-engine is built. The electrical engineers of to-day have taken this to heart very thoroughly, for their education and training requires as good a knowledge of mechanics as of electricity.

It may in truth be stated that the latter half of the nineteenth century will be famous as " The Electric Age," and will mark the development and application of a new and mysterious force, which has been tamed and harnessed through the indomitable energy of man to a thousand and one uses, benefiting all humanity. A force which does its work silently and swiftly, yet giving no sign of its presence ; a force subject to the greatest elasticity in division and subdivision ; and a force which in process of time will probably be called upon to be the universal slave of mankind, ready at a moment's notice to do his bidding, by land or sea, in the toiling city, or in the solitude of nature— another secret wrested from nature, signalising another triumph of mind over matter.

The repeal of the Electric Lighting Act of 1882, which was chiefly brought about by the efforts of Lord Thurlow in 1888, marked the beginning of a new and bright era for the electrical industry. Hitherto, its progress was terribly burdened by the old Act; this Act was hastily passed and put into force by Parliament in 1882, during the electric light mania, and was intended to protect the public against the wild and indiscriminate investment of money in swindling companies that were floated by the dozen. The protection took the form of empowering the local authorities to purchase the plant of the enterprising company at the end of 21 years, whether the company were agreeable or not, without being bound to pay the latter anything for the goodwill or source of profit, or increase of value, of the undertaking. But the bane of the Act was soon felt, for it totally crippled the electric light industry, so far as public supply was concerned. The golden stream ebbed back from an industry that was made so doubtful of remuneration, and electric lighting was starved through want of capital. Yet it is said that "out of evil comes forth good," but the only good apparent is that whereas a few years ago towns might have had poor and ill-designed plant put down, now electric plant and apparatus have reached a high state of perfection, the light and the system of distribution are better understood, and, lastly, the advent of the high-tension alternating-current system has supplied a wonderfully elastic means of distributing light and power over large areas—a problem which cannot be well solved by low-tension currents. The repeal of the Act in 1888 increased the term of compulsory purchase to a period of 42 years, thus doubling the length of time; this was the chief alteration, and it was one of incalculable value to the industry. The immediate effect of this acted as a great fillip, and the dormant energy that had been extant soon showed itself in subsequent years. The progress of the electrical trade has been phenomenal, and the basis of a new and important industry has been firmly established on sound foundations.

Electricity in England and Abroad.

Let us look around and see what position electricity occupies, and what progress it has made in various parts of the world; no account will be taken of the telegraphic or telephonic industries, but only the utilisation of electricity for lighting and motive purposes.

America takes the first place in the electric light and power industry, the towns there being lighted by the score—and there are very few towns indeed that do not possess a central electric light and power station for distributing same—the electric light over there being as prevalent as gas. They adopt a better standard of light, but we in England are accustomed to have a few dim gas lamps every 40 or 60 yards, which just suffice to make darkness visible. Probably the Britisher thinks it is a vast improvement to have light of any description in the streets of a town, for a century ago towns had no light scarcely, except what was eked out by the dismal oil lamp. An American reckons a good street light as valuable as a policeman, and no doubt he is right.

Coming now to the application of electricity as motive power. In the United States there are hundreds of tram lines worked by electricity, having a total length of track of several thousand miles, and employing close on 10,000 motor cars. In the space of one year it is calculated that between 400 and 500 million passengers are carried, and that the aggregate number of miles traversed is nearly 100,000,000, the amount of capital invested being 100,000,000 dollars, or £20,000,000.

The figures given in Tabulation 1 show the extraordinary development of electric tramways in that country, The statistics refer only to one system or method of working the lines, that known as the "Overhead System" because the car receives its current by a trolley wheel rubbing along a wire suspended over the top of the car. This is the system mostly adopted in the States, although there are a number of other lines worked on other systems, such as by accumulators, etc., and of which no figures are given.

TABULATION 1.

Year.	Lines.	Motor cars.
1884	1	—
1885	3	13
1886	5	39
1887	7	81
1888	32	265
1889	104	965
1890	126	2,000
1891	405	5,100
1892	550	7,500
1893	750	10,000

By far the greater number of the lines operated are run by two systems, the " Thomson-Houston " and " Sprague's"; the rest is run by several minor companies.

The greatest tramway, or street railway, system in the world exists at West End, Boston, Mass., originally worked throughout by horse power. Electricity is fast superseding horseflesh as a motive power, and very soon it is hoped that the entire mileage will be worked by electricity.

A great problem now occupying the attention of electricians and engineers is the utilisation of the Niagara Falls, millions of horse-power there running to waste year after year. Hitherto this vast source of power has been unavailable, and its utilisation to produce electricity will be one of the greatest practical achievements in the domains of electrical science. It is proposed to transmit the electricity generated to distant towns, there to be employed.

in lighting the streets and houses, driving machinery, propelling street cars, etc.

Turning our attention towards Great Britain, we cannot but be struck by the disparity existing between it and America. London is the only place where electricity enters to any degree into the commercial and professional world ; even in the largest city in the world, the heart and core of business and finance, the light is confined to the wealthy classes, so far as private dwellings are concerned, although it is freely used now in theatres, restaurants, hotels, clubs, etc. Yet it must be admitted that enormous advances have been made in London during the last few years. Every district in the metropolitan area is allotted out to some electric light company, and, furthermore, every district has now a central station put down and at work. Whatever the English do, they do thoroughly and solidly, choosing their own time and weighing everything carefully before committing themselves—and as the individual is, so is that larger individual, the Corporation or Local Authority—but when once on the move, nothing but the most substantial and approved work will suffice.

A short time back the towns possessing central stations could almost be counted on the fingers. A great wave of electrical activity has now begun to pass over England, and city after city, town after town, is up and doing. This activity among Local Authorities, etc., has broken out suddenly, like an epidemic, and it will not be very long before England leaves America far behind—if comparison be made on the score of size.

The lighting of the City of London was a great and decisive step towards showing that the time had arrived for Local Authorities to seriously consider the advisability of adopting the electric light as a better means of illumination. The lighting by gas in the City gave an aggregate of

35,000 c.p., the ordinary jets used being about 14 c.p. to 16 c.p., while the larger jets gave about 40 c.p. to 50 c.p., five cubic feet per hour being consumed by the small jet, and 15 cubic feet per hour by the larger jet. The above quantity of light has now been replaced by arc lamps of 1,000 actual candle-power in the main thoroughfares, and by 'incandescent lamps of 25 c.p. to 50 c.p. in the side streets, alleys, etc. The cost of lighting by electricity is almost double that of gas, but it must be remembered that in the former case much more light is given. The undertaking is in the hands of a company which was formed to take over the concessions granted to two eminent electrical firms—namely, the Laing, Wharton, and Down Construction Syndicate, and the Brush Company; the former using their Thomson-Houston system, and the latter the Brush system. Besides the street-lighting, a compulsory area is mapped out, and mains for private house supply are already laid down in this fixed area. The private supply is by high-tension alternating currents, this method being adopted by both the above contracting firms on their respective systems. The City is divided into three districts; of these the Thomson-Houston system is used in the Eastern District, which is east of the Mansion House, Princes-street, and Moorgate-street, while the Brush system is used in the Western and Central Districts.

In London, the Vestry of Saint Pancras has the honour of being the first Local Authority to adopt the electric light in the streets in place of gas, and also the first to erect a central electric light and power station, supplying current both to their own street electric lamps and to the private houses. The streets are lit by arc lamps of 1,000 actual candle-power, of the Brockie-Pell type, and placed at distances varying from 160ft. to 245ft. apart, suspended on cast-iron poles 25ft. high. The private supply to householders was met by having machinery capable of running 10,000 glow lamps of 16 c.p. each, the whole output of which was taken up within six months after the erection of the station, and since then the plant has had to be increased considerably.

The following tabulation shows the number of lights used in London corrected to end of 1893 :

TABULATION 2.

Company.	Lamps in use, 16 c.p.
London Electric Supply Corporation	33,000
City of London Electric Light	33,000
Westminster Electric Light (three stations)	66,000
House-to-House.....	14,000
Chelsea Electric Light................. -	20,000
St. Pancras Vestry	15,000
Kensington and Knightsbridge (three stations)	32,000
St. James and Pall Mall (two stations) for one	30,000
Metropolitan Electric Light (five stations)..	82,000
Charing Cross and Strand Electricity Supply Corporation	18,000
Notting Hill Electric Light	6,000
Total.....	349,000

The demand for current in some districts for a long time past has greatly exceeded the supply, as the various companies cannot extend their mains quickly enough. Some house-holders have had their houses ready wired for months, and have had to wait until the supply company's mains get laid near their door. Great complaints naturally arise from this ; sometimes it is the fault of the supply company through backward state of work, but mostly it is due to the importunate solicitations of the house-wiring tout, who, to secure the order for wiring, victimises his customers by assuring them with the most bare-faced plausibility that the supply mains will be down at his door by the time the house is ready. The householder is, therefore, strongly advised to take timely warning. Before putting a house into the hands

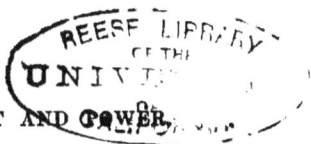

of any stray individual, first see that the supply mains are already laid down, or being laid down close by, or, at all events, within such a distance that the supply company, upon being asked, will join them to the house mains, then, and not until then, should one have his house wired.

In the provinces there is a steady increase going on which is very satisfactory. Numbers of our large towns have spent large sums of money in establishing Central Electric Light and Power Stations.

Out of the 50 towns that possessed central stations by the end of 1893, 35 are the property of private companies, the remaining 15 being established by corporations. It is interesting to note the different methods of distribution adopted : 29 towns use the high-pressure alternating system, 11 the three-wire low pressure, one the five-wire low pressure, and nine the simple two-wire low pressure. The solitary five-wire system is adopted by Manchester ; and the result of it will be awaited with interest. Three different driving methods are used—15 using ropes, 15 using belts, and 19 having direct-driving engines. The tendency, however, in English central-station work is to have direct-driving high-speed engines, coupled to low-speed armatures. Nearly all the American stations use shafting and belts, and so take up a considerable floor space. But it is being found that the shafting and belts, with their numerous pulleys and friction clutches, require a lot of looking after, and there are indications that American engineers will prefer to follow the English tendency and use direct-driving plant. It is mostly the earlier stations that are using rope and belt driving, and mostly the later stations that have adopted direct driving. With shafting, one or two powerful engines drive all the dynamos, so that if anything happened to the engine where there is only one, all the dynamos are stopped. With a direct-driving plant, an accident to an engine limits its effect to its

own dynamo. Central-station plant is getting now fairly
stereotyped in character, particularly as the bulk of the work
done is in the hands of half-a-dozen or so large contractors.
Designers are thus beginning to follow certain lines, and as
time goes on the general practice will become more and more
stereotyped. It will be the same with regard to the distribution of the electricity. Experience obtained from towns
differing widely in their characteristics and requirements will
most probably point to certain evolved systems as being the
most suitable for certain cases. So that eventually there
will arrive a time when a consulting engineer will have to
follow more or less well-prescribed plans. In order to arrive
at this state of things it is unfortunately necessary to pass
through the " weeding out " stage ; all manner of central
stations have been erected in England, most of which show
the mark of the prevailing idea of the engineers who designed
them ; those which prove to be well designed, not only for
present use, but for future extensions, will repay well the
enterprise of the company or corporation who erected them,
and who will have cause to congratulate themselves; whilst
those which prove to be badly designed, particularly with
regard to extensions, will have a sorry time of it, because
it entails great expense to modify a system of distribution,
and if not modified, things would go from bad to worse.

Passing on from lighting to traction, until within the last
year or two electric traction in England was very backward,
there only being two or three small tram lines which were
worked by electricity—such as at Blackpool and Brighton,
and one or two other places. The year 1890 may be taken
as the era which saw the introduction of electric traction in
England on a large and thoroughly commercial scale, for at
the close of that year the City and South London Electric
Railway was opened to the public ; and what makes this
event of double importance is that it is the first actual

railway in the world to be operated throughout by electric instead of steam locomotives, all previous electric traction being confined to tram lines or, at most, to light street railways. Hence, although behind America a long way with regard to the application of electric power for traction purposes, yet in heavy and large engineering work England acts as pioneer.

One year afterwards (in November, 1891) the Roundhay Electric Tramway was opened at Leeds, this being the first tramway in England operated on the overhead trolley system. This undertaking was on the Thomson-Houston system, the total equipment—cars, machinery, etc.—being imported from America, even to the steam-engines. Everything supplied being cut and dried, and the same as erected in several hundreds of towns, the system from the start ran perfectly, without any of the experimenting and trials which, so far, have been so prevalent a feature in this country in most of the attempts to work tram lines by electricity. The next tramway that claims notice is that constructed in South Staffordshire, and opened in January, 1893. This was also on the overhead trolley system, but designed differently to the one at Leeds; the chief difference being that while at Leeds the trolley wire was suspended over the centre of the track by span wires, in the Staffordshire line the trolley wire was supported by light brackets which projected from poles fixed on one side of the road only, thus doing away with span wires, which to some people form the objectionable part of the overhead system.

The last undertaking respecting electric traction is the Liverpool Overhead Electric Railway, which was opened for traffic in February, 1893. From the experience gained in the construction and subsequent working of the City and South London Railway, several improvements were able to be introduced in the construction of the Liverpool railway, and

as time goes on those which follow will become more and
more improved, and the undertakings will become larger and
more ambitious.   Already there are several electric railway
schemes prepared for the Metropolis, and the construction of
tubular underground electric railways promises to afford a
satisfactory method of relieving the enormous and congested
traffic of large towns.   With regard to the employment of
electric instead of steam locomotives on our main lines, at
present conjectures only can be made ; there is no doubt it
will eventually occur, but it will take time, and the change
will be gradual.

Advantages of the Electric Light.

A pure and healthy atmosphere is greatly dependent upon
the amount of oxygen present, and the following are the per-
centages of gases which, being mechanically mixed, form the
atmosphere :

Oxygen.....	23 per cent.	by weight.
Nitrogen	77 ,,	
Total	100	

There are, in addition, very small quantities of carbonic acid
gas (CO_2), hydrogen, ammonia, and organic matter.

All substances when burning consume oxygen—that
is to say, the act of combustion of matter signifies the
chemical combining or mixing of the vapours of the burning
matter with the oxygen of the atmosphere.   Hence if a substance be burnt in a closed vessel containing atmospheric
air, it would continue to burn until all the oxygen of the
air had been abstracted, and when that stage had been
arrived at, the combustion would cease through lack of
oxygen.   Some of the gases thus given off are often poisonous,
according to the nature of the article burnt.   The most
common ways of obtaining light is by oil, candles, or gas.
Of these three, candles are the most unhealthy light, because

they vitiate the air more than the other methods of illumination ; all three vitiate the atmosphere and give off poisonous fumes, some of the carbon which they contain mixing with the oxygen from the air, and so producing the deadly gas known to chemists as carbon monoxide, or CO. The operation of breathing has a similar effect on the atmosphere— pure air is inhaled by the lungs, but impure air, containing carbon monoxide gas, is exhaled. It is this that makes a room feel close or stuffy when a person has been sitting in it for any length of time with doors and windows closed. One gas jet of 14 c.p. consumes about 20 cubic feet of air per hour. A man with good respiration will consume about four cubic feet of air. From this it will be seen that one gas jet vitiates a room to the same extent as four persons would do.

The following table shows forcibly the weight of the preceding remarks, each illuminant giving a light of 12 standard candles for one hour :

TABULATION 3.

Illuminant.	Cubic feet of carbonic gas produced.	Cubic feet of air vitiated.	Heat evolved in lbs. of water raised 10° F.
Electric light	None.	None	$13\frac{3}{4}$
Common gas.........	$3\frac{1}{4}$	$348\frac{1}{4}$	$278\frac{1}{4}$
Paraffin oil............	$4\frac{1}{2}$	484	362
Wax candles.........	6	$632\frac{1}{4}$	383

Illumination by means of the electric light keeps the atmosphere pure and cool. Respecting the first claim— namely, that of purity—the proof is made self-evident by simply stating that the light is enclosed in an air-exhausted and hermetically-closed glass bulb, thus absolutely severing the light from any contact with the atmosphere. Under

c

these conditions the lamp filament cannot burn, through absence of oxygen. Most people fancy that the electric light does not give out heat ; this is a very great mistake, since there can be no light without heat. When a hydrocarbon, like coal gas, burns, its carbon unites with the oxygen of the atmosphere, and this chemical affinity causes such a great evolution of heat that the solid carbon particles are raised to a state of incandescence, or white heat ; so we obtain light.

In the case of an electric incandescent lamp, a small amount of heat is evolved from it. This is shown by taking in the hand a lamp that has been burning some hours when it will be found that there is a comfortable degree of warmth without burning the hand. Compare this with the effect of placing the hand close to a gas jet of equal candle-power, and a practical example is found at once of the relative heating properties of the two illuminants.

The great value of the electric light and the boon it confers cannot be over-estimated, particularly when it is introduced into manufactories, mills, shops, etc., where the articles lying about are of a highly inflammable nature, and the danger of fire so great in the presence of naked gas lights. The act of lighting by electricity can be done automatically and instantaneously throughout the whole of a large establishment ; this avoids the risks and dangers of gas lighting by hand, as well as time and labour, and therefore expense in so doing.

The best proof that can be put forward as to the great safety in the use of the electric light, is that the insurance companies look upon its introduction with favour, when properly installed, yet it must be borne in mind that the insurance people have had to draw up very drastic rules in order to protect themselves and the public from the bad work of incompetent persons ; the electrical industry should be greatly indebted to them on this score, because without any such compulsion unscrupulous firms would scamp their work, thus bringing discredit upon the whole industry. These kind of people never hesitate to do electrical work at prices with which a firm putting in good work cannot possibly compete. Following in the wake of this type of contractors comes a long train of nondescript individuals, who take to themselves the title of electricians, electrical engineers, or anything else to the purpose that suits their fancy. These enterprising gentry recruit their ranks from tinkers, plumbers, ironmongers, bellhangers, and such like, and trade upon the ignorance of the public. This is one of the attendant evils that dog the progress of a new industry—a veritable "old man of the mountain," and the same thing occurs in the development of every industry. Time only can cure it.

Before concluding these remarks, there should be pointed out one other advantage which the electric light possesses, and a matter of paramount importance—namely, health and personal comfort. Who is there who has not felt the exhausting and injurious effects of a heated and contaminated atmosphere? The stifling feeling in a theatre, or any other place where there is a number of people, is partly due to themselves and partly to the gas lights. The same applies to workrooms of every description. That a substantial improvement is wrought by the introduction of the electric light there is no shadow of a doubt. In the General Post Office, St. Martin's-le-Grand, the electric light mostly replaces gas, hence a healthier atmosphere is procured ; and the great benefits and comforts which it bestows on the employés are such that it has diminished the hours of absence two hours per day. In addition to this, the staff are able to work better and quicker, due to the absence of the lassitude so often felt in heated atmospheres. Mr. Preece, the Chief Electrician,. considers the increased amount of work obtained from the staff very nearly pays for the cost of lighting.

In conclusion, it may be said that *the electric light is the safest of all illuminants, provided that the work of instalment is well and conscientiously carried out.*

Cost of Electricity v. Cost of Gas.

A great deal of misleading and incorrect matter has been published concerning the relative cost of gas and electricity. Some writers have made the cost of electricity two to three times that of gas ; others, again, have erred in the opposite direction, and tried hard to show conclusively that it is quite as cheap as, if not cheaper than, gas. The author is of opinion that neither party is correct. The cost of electricity depends a great deal on the circumstances and conditions of production. To show clearly what is meant by this, take the case of a large mill owner, who has, say, a 200-h.p. steam engine ; from this, 20 h.p. can easily be reserved for driving a dynamo, very little more coal will be burnt, scarcely any attention required, the only outlay of money being for the dynamo and fixing up of lamps—here the cost of the light will be far below gas, possibly not much more than one-half. Now, take the other extreme, buying current from an electric supply company in London ; the general charge is 7¼d. per unit, and this works out to an equivalent cost of 6s. 4d. per 1,000 cubic feet of gas, which is more than double the average cost of gas at 3s. per 1,000 cubic feet. From these two examples the reader will at once perceive how absurd it is to give a sweeping solution of a problem which admits of two such diverse interpretations. A detailed comparison between gas and electricity will now be made—assuming that electricity is supplied from a central electric light station, just as gas is supplied from gas works, for this is the way the general public will obtain their electricity.

(a) Comparison with an Incandescent Lamp.—An ordinary-sized gas jet such as is found in all private houses may be judged to give out 14 actual candle-power, when in good condition, the consumption of gas being taken at the rate of five cubic feet per hour, the gas being of good quality and at a normal pressure. With gas at 3s. per 1,000 cubic feet, the cost of five cubic feet would be 0·18 penny, so that one gas jet of 14 c.p. burning for the space of one hour would cost 0·18 penny, or little less than one-fifth of a penny, or one penny if burning for five hours ; and for 1 c.p. the cost would be 0·013 penny or one seventy-seventh $(\frac{1}{77})$ of a penny, or one penny if burning for 77 hours—1 c.p. being taken as the standard measure of light as given by a pure spermaceti wax candle which consumes 120 grains of wax per hour. Passing on now to electricity, assume the price of current is 7¼d. per Board of Trade unit, this being the charge made by the Metropolitan Supply Company in London, and is a fair average price. A unit signifies a certain amount of electrical energy (1,000 watts) acting for the duration of one hour. An incandescent lamp of 16 actual candle-power requires 54 watts, or 3·5 watts per candle-power ; this works out at 0·025 penny, or $\frac{1}{40}$ penny, per hour for 1 c.p. Besides this cost there is the cost of the lamp itself to be considered. An incandescent lamp costs 3s. 9d., and on an average has a life of 1,000 hours, at the end of which time the filament breaks, or the lamp becomes so blackened that its light-capacity is worthless. Therefore, the cost of the lamp comes to ·0028 penny, or one three hundred and fifty-seventh of a penny $(\frac{1}{357}$ penny) per hour of candle-power. Adding these two costs together we have

Cost of current .... = ·0250 penny
Cost of lamp .... = ·0028 „

Total .... = ·0278 penny.

Hence $\left\{\begin{array}{l} 1 \text{ c p. per hour from gas} \\ 1 \;\; „ \quad\quad „ \quad\quad „ \quad\text{ electricity} \end{array}\right.$ = ·013 penny.
= ·0278 „

or $\left\{\begin{array}{l} 14 \;\; „ \quad „ \quad „ \quad\text{ gas jet} \\ 16 \;\; „ \quad „ \quad „ \quad\text{ glow lamp} \end{array}\right.$ = ·18
= ·44 „

Gas being reckoned at 3s. per 1,000 cubic feet, and electricity at $7\frac{1}{4}$d. per 1,000 watt-hours. ·44 penny means a trifle under one halfpenny, so that in practice it is usual to take the running cost of a 16-c.p. incandescent lamp at one halfpenny per hour. Similarly, the cost of an 8-c.p. lamp is put down at one farthing. Of course this only applies when the cost per unit = $7\frac{1}{4}$d. If the cost be less than $7\frac{1}{4}$d., say 6d., which is what the St. Pancras Vestry charge, then the cost of a 16-c.p. lamp per hour would be $\frac{1}{2} \times \frac{6}{7\frac{1}{4}} = \frac{3}{7}$ penny nearly. When the number of lamps used and the hours that they are burning are a fixed number, say in a business establishment, it is very easy to arrive at the total cost of the light. Suppose 50 lamps of 16 c.p. are lit at dusk every day, except Sundays, until 8 p.m., upon referring to the table of the lighting hours given at the end of the book, it is seen that the number is 742 ; as Sundays are excluded, one-seventh of this $(\frac{742}{7} = 106)$ must be deducted, leaving $742 - 106 = 636$ hours of lighting, so that $\frac{1}{2} \times 50 \times 636 = 1,990$ pence = £66. 5s., using electricity at $7\frac{1}{4}$d. per unit.

From the figures given above, it is seen that electricity in the form of incandescent lamps costs more than double gas, since ·0278 is more than twice ·013, therefore the equivalent cost of gas to bring the price up to that of electricity must be—

$$\frac{3 \text{ shillings} \times \cdot0278}{\cdot013} = 76 \text{ pence} = 6\text{s. }4\text{d.}$$

The table below is taken from a paper written by the author, entitled "Shall Gas Undertakings Supply Electricity?" (see *Transactions* of the Incorporated Gas Institute, 1890), and is worked out to give the equivalent cost of gas for different prices per unit of electricity, the figures here being corrected ones :

TABULATION 4.

Pence.

Unit of electricity	4	4½	5	5½	6	6½	7	7½
1,000 cubic feet of gas	42	48	58	58	64	69	74	79

Unit of electricity	8	8½	9	9½	10	10½	11	11½	12
1,000 cubic feet of gas	85	90	96	101	106	112	117	123	128

The above figures relate only to lighting by means of incandescent or glow lamps, and not to arc lamps. It will be observed that by multiplying the price per unit of electricity by the number 10½, we obtain the corresponding equivalent cost per 1,000 cubic feet of gas. For example, let 5½d. be the cost of a unit, then 5½ × 10½ = 58, and on referring to the table it will be seen that 58 is given as the equivalent cost of gas.

(b) Comparison with an Arc Lamp.—When the greatest economy is required, arc lamps should be used ; this kind of electric light is particularly suitable for large spaces, such as building yards, warehouses, etc. A 1,200 nominal candle-power arc lamp takes about 330 watts, or one-third of a unit, and may be said to give out about 480 actual candle-power when a thin ground-glass globe is used. Wit electricity at 7¼d. per unit this works out to ·0053 penny

per candle-power per hour. There is now to be added the cost of the carbon consumed in the lamp. This may be put down at ¾d. per hour, therefore this equals ·00125 penny per candle-power per hour. Adding both together, we have—

Cost of current...................... = ·0053   penny

,,   ,, carbon .... = ·00125   ,,

Total .... = ·00655   ,,

Hence an arc lamp giving 480 actual candle-power costs 3d. per hour.

It was found that the cost of gas at 3s. per 1,000 cubic feet was ·013 penny per candle-power per hour—hence the equivalent cost of gas is

$$\frac{3 \text{ shillings} \times ·00655}{·013} = 18·1 \text{ pence, or little more than 1s. 6d.}$$

From this it appears that an arc lamp of 480 actual candle-power, when using a thin ground-glass globe, only costs about one-half the price of an equal amount of light from gas. This is taking gas at 3s. per 1,000 cubic feet and electricity at 7¼d. per unit. With these same prices for gas and electricity, it was shown that the cost of an incandescent lamp is more than double that for an equal amount of light from gas; hence an arc lamp costs less than one-fourth the cost of an incandescent lamp for equal candle-power. In the case of incandescent lamps, the equivalent cost for gas was obtained by multiplying the price per unit by 10½, and since arc lamps cost less than one-quarter of incandescent lamps, multiplying the price per unit by 2½ gives the equivalent cost for gas in the case of arc lamps.

The tabulation below shows in a brief form the respective costs per candle-power per hour for arc lamp, incandescent lamp, and gas jet; also the cost of their respective full candle-power, and finally the ratio of their cost, taking gas as a standard at 100.

TABULATION 5.

Illuminant.	Cost of 1 c.p. per hour.	Full c.p. per hour.	Ratio.
Arc lamp......	·00655 or $\frac{1}{152}$ penny.	For 480 c.p. = 3 pence	51
Incandescent	·027 or $\frac{1}{36}$,,	,, 16 c.p. = $\frac{1}{2}$ penny	207
Gas jet........	·013 or $\frac{1}{77}$,,	,, 14 c.p. = $\frac{1}{6}$,,	100

Take an establishment lit by arc lamps, say 10 in number, and running the same hours as was given in the example with incandescent lamps, that is 636, excluding Sundays. At 3d. per lamp per hour, this $= \dfrac{636 \times 3 \times 10}{20 \times 12} = £79.$ 10s., and is the cost per annum for illuminating power equal to $10 \times 480 = 4{,}800$ candles, or about 342 5ft. gas jets.

All the above calculations have been based on the standard size of lamp, or that size most commonly used—that is, 16 actual candle-power for incandescent lamps, and 1,200 nominal, or 480 actual, for arc lamps As the lamps, both incandescent and arc, increase in candle-power or size, so the electrical power consumed by them becomes slightly less in proportion to the candle-power. Hence, their cost will become slightly less also, but against this advantage it must be remembered there is a greater concentration of light, which is a disadvantage. This matter will be discussed fully further on.

Economics of the Electric Light.

So far nothing has been said concerning the numerous small savings that electricity effects. The foregoing detailed comparison of cost simply refers to direct charges. Upon examination, we shall find that the direct charges, or what

may be termed "prime calculated cost," will be considerably
altered in value. Take into account the great waste of gas that
occurs through lights being left burning unnecessarily. How
many times a day are gas jets left full on upon the occupant
leaving a room for, say, 10 minutes? It is too much trouble
to turn the gas out, and particularly to relight it; matches
are not always at hand, and groping about in the dark is
very unpleasant—a feat more often than not attended with
knocking one's head or shins against articles anything but
soft. Turning out the gas or lowering it may seem to effect
a very trifling saving, but it is these trifles that soon add
up into a respectable total—"Many a mickle makes a
muckle." Three times 10 minutes every day signifies close
on a couple of hundred hours per annum. Now perform this
operation on half-a-dozen gas jets or incandescent lamps; it
reaches the total of nearly 1,100 jet or lamp hours, or equivalent to one jet or lamp burning for 1,100 hours. In the case
of gas this signifies nearly a sovereign wasted, and in the case
of electricity more than a couple of sovereigns saved. Where
electricity is employed as a lighting medium there is not the
slightest excuse for such needless waste, because by fixing
the switch close by the entrance door the light can be
immediately switched on before entering the room; similarly
when leaving the room, the light need not be switched off
until just at the door. Only those who have experienced it
can appreciate the comfort of the electric light in a bedroom,
with the switch fixed close to the bed head, and ready to
light up or turn out in an instant. Another matter of considerable importance is the great saving that electricity
ensures with regard to expensive decorations, whether
metal work, paint, curtains, and draperies of delicate hue,
moulding, statuary, and the thousand and one beautiful
things that adorn a handsome hall or room. Such places
as theatres, hotels, public buildings, etc., recognise this,

and it is common knowledge that some of the highly-decorated places of amusement save a very heavy sum annually, which inevitably must have been spent had gas been the illuminant. Whilst speaking about amusement places, hotels, etc., it may readily be inferred that those which possess the cool atmosphere due to electric lighting are more likely to, and, in fact, do, obtain a greater share of patronage than the others, other things, of course, being equal. Each reader can now place his own figure of value on the foregoing economic features of electricity, and this, deducted from the " prime calculated cost" of electricity, will give a fair idea of the "true absolute cost " of the electric light as compared with gas.

Sources of Power.

(a) *Artificial.*—People look considerably surprised when they are informed that the production of electricity, and hence the electric light, is attendant with the consumption of a good deal of power ; that, for example, a steam engine yielding six brake horse-power will only run 60 incandescent lamps of 16 c.p. at the utmost. People know that there is such a thing as a " battery," or a " cell," and have a dim recollection that zinc or some other metal is put into a jar containing some acid, or chemical solution, and that from this simple apparatus one can obtain an electric light—in a fashion—so when they learn that a boiler and steam engine is required to run 50 or so incandescent lamps, or half-a-dozen arc lamps, they cannot understand the reason. It is interesting to draw a comparison between the power generated by a battery and by a dynamo. For every pound of coal burnt per hour under the boiler, $\frac{1}{3}$ b.h.p. can be produced, and this would run three incandescent lamps of 16 c.p. each ; if a battery were employed to supply the same amount of electrical energy for the same time, it would be found that $\frac{1}{4}$lb. of zinc would have to be burnt-

The cost of the coal may be put down at less than $\frac{1}{10}$th of a penny, whereas the cost of the zinc is twopence, or the fuel for a battery is 20 times more expensive than the fuel for a boiler and steam engine.   This is why batteries are entirely out of the question for generating large quantities of electrical energy ; and even if the cost of the battery fuel was much less, it is doubtful whether they could be of much practical use on a large scale.   The chief way to obtain power is by burning some material either in a solid, liquid, or gaseous state.   As a solid, coal is burnt under a boiler, and the heat so obtained utilised in converting water into steam, at a greater or less pressure.   As a liquid, petroleum or other oils can be cited, the spray of which being mixed with air is exploded, the explosion yielding energy, arising from the expansive force of the gas.   As a gas, coal gas is compressed and exploded.   Hence, the steam engine, oil engine, and gas engine, using respectively a solid, a liquid, and a gaseous fuel, belong to the class of artificial sources of power, and bear witness to the ingenuity of man.

(b) Natural.—Besides these, there are the great natural sources of power which engineers are continually directing and subjecting to their use—a mountain torrent, a swift river, a waterfall, wherever a body of water is moving from a higher to a lower level—all will yield a continuous supply of energy, in great or small quantities, according to the rapidity and volume of water rushing down.   Then there is the ebb and flow of the tide, which is the greatest source of wasted energy on the globe.   Twice daily do the billows surge and thunder on the shore, and it is only when one gazes fascinated at the wide glittering expanse of seething waters as they hurl themselves on a rock-bound coast, that any slight idea can be formed of the terrific forces at work, millions of horse-power being wasted every mile of

coast they encircle. Truly, the foam-flakes, the expressive token of all this energy, possess as much value as if they were solid flakes of silver, could but this immense display of power be utilised! Attempts have already been made to store the force of the tidal motions, and, at the present rate of invention, we shall undoubtedly in the near future arrive at the ways and means of doing so. Another natural source of power is the wind, from the gentle breeze to the fierce tornado. This power has been used for driving ships for ages, and although steam has supplanted most sailing ships, still a good number remain. On land the many windmills on elevated ground scattered all over the country indicate how its power is utilised, although its services are here very fitful and erratic.

Conservation of Energy.

All energy on the face of the earth, partly excepting the tidal motions, has its origin in the sun. Coal is the product of the sun's rays acting countless ages ago on vegetable matter and the undergrowth of primeval forests, the heat of the sun thus being stored. By digging out the coal and burning it, we regain the heat which was dissipated in the remote past. In the vast and dense forests of Central Africa and other unexplored parts of the world, the sun is probably storing up for us untold wealth, ready for the future use of the denizens that will some day cultivate the land. It is the same with water power; it is the energy of the sun's rays which evaporates the water in our seas and lakes, and this water vapour, ascending into the sky, is wafted by the winds hither and thither until, coming into strata of cool temperature, the vapour condenses into a rain cloud, and returns to the earth's surface in the form of rain. When the rain falls on elevated regions, such as mountain summits, a mountain

torrent is produced, which as it dodges and dashes down to
the plains below mingles with other torrents, and so the
body of water accumulates until it meanders, a broad, shining
river, over the plains on its way to the sea. Thus the
cycle of changes go on, the water taken up into the clouds
from the sea being caused to return again. Again, it is the
sun's rays which produce winds : the heated air rises upwards,
and immediately a rush of cold air takes place to fill the
space vacated.

From the above general remarks it will be seen that really
there is no waste of energy. The word "waste" is only used
to express that which we cannot utilise for our own purposes,
hence there is only a transformation of energy going on from
one form to another. This great fundamental law of
nature is termed the "Conservation of Energy," and it must
be implicitly accepted, and thoroughly recognised and under
stood, when dealing with "electric energy," just as it should
be when dealing with any other subject which treats of
energy. To make this statement more emphatic, let us
follow the several steps of transformation of energy that
take place in reproducing the light of the sun in the form
of an electric lamp. First, the heat is stored as coal, which
has been very aptly termed by Mr. Preece "preserved sun-
beams"; coal being burnt, evaporates water into steam,
steam drives the steam engine, and thus this third transformation consists of changing the energy of heat into mechanical energy. By causing the steam engine to drive a dynamo
a fourth change occurs, the mechanical energy being now
converted into electrical energy. The final transformation
occurs when the electric current in passing through the lamp
filament generates such intense heat that it produces light.
Hence there are five transformations of energy between the
light of the sun and that of the electric lamp. It may be
asked, Why is it necessary to have so many changes, for

each change signifies a loss, according to the commercial efficiency of the machines or converters employed ? and, Why cannot we obtain electricity direct from coal, without these intervening converters ?   The answer is that our present state of knowledge will not permit us.   The problem has been attacked, but so far without the slightest practical value arising.   Five different transformations is a good number, and a roundabout way certainly, but as we see no other way, we must put up with it.

The following remarks give an idea of the losses that occur in each converter, at the present state of efficiency :

Of the total heat energy contained in burning coal, only 50 per cent., as a rule, can be said to be utilised to evaporate water into steam; hence the average boiler has an efficiency of only 50 per cent.

Passing on to the steam engine, it may be said that the very best steam engines only utilise from 18 to 20 per cent. of the heat energy supplied by the steam, and since the efficiency of a boiler is 50 per cent., the engine can be said to utilise only 9 to 10 per cent. of the heat energy evolved by the burning coal, allowing the low figure of 2lb. of coal per brake horse-power—a performance very rarely met with.   On allowing 4lb. of Welsh coal per brake horse-power, the efficiency of the engine is brought down to 9 or 10 per cent.—that is to say, that only 10 per cent. of the heat energy of the steam admitted into the cylinders is converted into available power, and when calculated from the heat energy of the coal burnt under the boiler, the amount utilised becomes 5 per cent.   However much boilers and steam engines may be improved it cannot, by any possible way, prevent this source of power from being terribly wasteful.   On theoretical grounds, the employment of fuel in the form of a gas (such as coal gas) and oils (as petroleum) to drive internal-combustion engines, known as gas engines and oil engines, gives a much higher efficiency than the best-made steam engines would, sometimes twice the efficiency ; but since this form of fuel is more expensive than coal, steam power is most economical for large power.

We have now reached the third converter—the dynamo, the machine that transforms the mechanical power given by the driving wheel or driving shaft of the engine into electrical power. Concerning the marvellous state of efficiency to which a modern dynamo has now been brought, it is difficult to find a parallel example. No less than 96 per cent. efficiency has been obtained in these machines for transforming mechanical power into electrical, and every dynamo on the market can point to 87 per cent., and, in addition to this, the small percentage of loss can be entirely accounted for and measured, thus proving up to the hilt the law of the conservation of energy. The fifth and last transformation takes place in the electric lamps, where the electrical current spends its power in heating up the carbon filament of the lamp to an incandescent state, and producing light. The energy that caused the sun to shed out light and heat thus, through numerous changes, reappearing in the light and heat of a miniature sun or electric lamp.

The commercial efficiencies of the several converters are tabulated below, taking 4lb. of coal for 1 b.h.p.

TABULATION 6.

—	Heat units.	B.H.P.	e	E
			%	%
4lb. of coal burnt per hour produce...	58,000	22·6	100	100
Amount of heat utilised by boiler......	29,000	11·3	50	50
,, ,, ,, ,, engine ...	2,563	1·0	8·8	4·4
,, energy ,, ,, dynamo..	2,383	·93	93	4·1
,, ,, ,, ,, lamps ...	2,330	·91	98	4

e denotes the commercial efficiency of each converter—

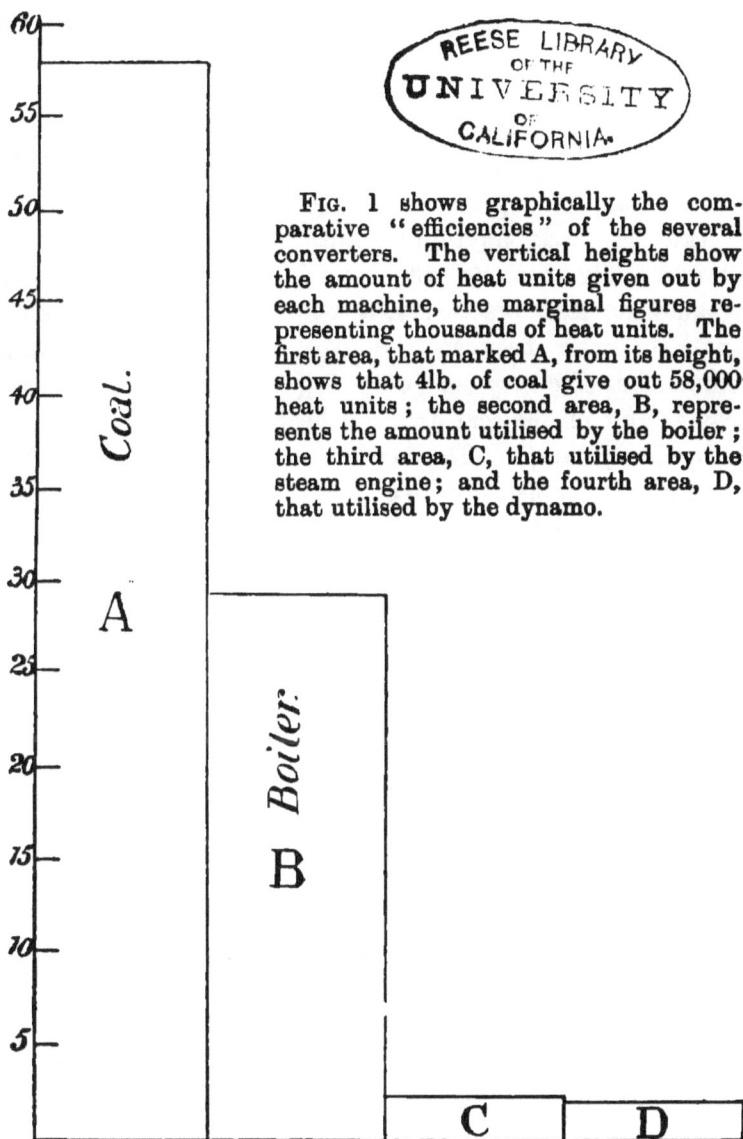

FIG. 1 shows graphically the comparative "efficiencies" of the several converters. The vertical heights show the amount of heat units given out by each machine, the marginal figures representing thousands of heat units. The first area, that marked A, from its height, shows that 4lb. of coal give out 58,000 heat units; the second area, B, represents the amount utilised by the boiler; the third area, C, that utilised by the steam engine; and the fourth area, D, that utilised by the dynamo.

FIG 1

D

that is, the amount of work done by the machine, as compared to the work put into the machine. E denotes the "commercial efficiency" of each converter as compared with initial energy supplied by the coal.

The last transformation—*i.e.*, converting the electricity into light—is given as with a loss of 2 per cent. This loss is entirely dependent on the distributing wires, and so can be made anything; but 2 per cent. has been allowed. By running the electric lamps close by the dynamo no loss, practically, is incurred, because there is no length of wire between the two. There is thus a total commercial efficiency of only 4 per cent. between the heat energy of the coal and the heat energy as it appears in the electric lamps. This is when steam power is used. No wonder that people strive to obtain electric power direct from burning coal, without the intervention of the wasteful steam engine.

CHAPTER II.

Coal as Fuel—Work and Horse-power—Mechanical Equivalent of Heat—Plotting Curves—Steam Boilers, Engines, etc.—Running Cost of Steam Engines—The Gas Engine—Running Cost of Gas Engines—Dowson and other Gas Producers—Petroleum Oil Engines—Water Power.

Coal as Fuel.

It is the sun that, through its rays of heat energy, causes trees, plants, and other vegetable and organic matter to spring out of the earth, grow, flourish for their allotted time, afterwards to gradually decay and die off; decomposition then sets in, and the matter sinks into the earth whence it came. The unceasing action of the sun's energy bring forth a fresh supply of living vegetable matter to take the place of the dead; this in its turn decays and dies after its season, and so layer after layer gets buried in the earth, each layer passing through successive chemical changes, due partly to the internal heat of the earth, etc. In this way coal is formed. The existing coalfields have been formed from decomposed plants and herbaceous matter that flourished as great primeval forests in ages so remote that the lapse of time may be counted as millions of years. At that early period of the earth's division into land and water, we have every reason to imagine the soil entirely covered with a dense and luxuriant growth of ferns, reeds, grass, etc., of a gigantic size, the soil being partly submerged in water,

D 2

thus forming huge swamps. This extraordinary carboni-
ferous age is attributed to the humidity of the soil and the
warm or possible tropical climate that was likely to exist at
that time in places which are now temperate or frigid. No
human being or animal could have lived in this Vegetable
World in the midst of vapours evolved by the decomposing
matter, the only living objects that inhabited the earth
being, so far as we can tell, curious forms of reptile-fish
and such like imperfect organisms.

It is easy to see how the heat energy of the sun is
transferred into coal, in the form of latent or stored heat,
which ages after becomes liberated upon the coal being
burnt. This provides one of the most beautiful examples
of the conservation of energy. Living plants and vegetable matter derive their food from the atmosphere,
absorbing the carbonic acid gas, or carbon dioxide (CO_2),
that is present. The sun expends part of its heat rays in
decomposing this gas—that is, its warmth acting on the
delicate leaf cells, has the effect of splitting up the carbon
dioxide into its component parts—namely, carbon and
oxygen—so that every molecule of CO_2 is split up into
one atom of carbon and two atoms of oxygen. Now, to
tear away the oxygen atoms from the carbon atom, or, in
chemical parlance, to overcome their chemical affinity, requires
a great amount of energy, and since energy is a form of heat,
a great deal of heat must be expended by the sun to do
this. The oxygen atoms being set free return to the atmosphere, the carbon atoms remain behind to form the structure of
the plant; this is why all woody matter, etc., consists greatly
of carbon. The first opportunity that occurs, the carbon
atoms join on to oxygen atoms, so as to regain their condition as CO_2. When oxygen and carbon combine they
always evolve great heat, so when coal or any other car-
boniferous matter is burnt in the atmosphere, the carbon

seizes hold of the oxygen in the atmosphere, and these
combining give out such heat that the carbon atom
is made incandescent, or white hot ; this action being continuous, produces what we call flame. The flame of a candle
is a hollow cone, the cone being formed by a thin film or
envelope of glowing carbon. When the air supply becomes
scant, the candle grows dim, simply because there is not sufficient oxygen present to combine with the carbon, consequently the latter only get partially consumed, and smoke
is given off instead of a glow vapour. From this it can
be understood that the heat given out by burning coal is
exactly equivalent in amount to the heat that was expended
in producing the coal. Returning, however, to the formation
of coal, the first result of the decomposition of the vegetable
matter is a produce called peat, a dark brown solid soil ; the
richer kinds being almost black. This peaty soil is found
extensively in Ireland, in the swamps, marshes, and bogs
with which that country abounds. The depth of the soil
varies from a couple of feet to 50ft. ; the top layers are
usually of a light reddish-brown colour, deeper down the
hue changes to a dark brown, and still deeper it is almost
black. As the depth increases, the soil, in addition, gets
heavier and more solid, so that the surface peat is light and
fibrous, and the black peat heavy and compact. It is used
by the peasantry for fuel, after being first well dried. When
thoroughly compressed and dried, it gives an evaporative
power when burnt equal to one-half that given by ordinary
coal, comparing equal weights. It is a great deal used as
boiler fuel in districts where it is obtained in quantity. The
second stage produces "bituminous or common coal"; this
after a while becomes harder and harder until the third
stage is reached, when "anthracite coal" is produced. Any
further lapse of time turns this kind of coal, which is the
best, into fossilised coal, which is so hard that it is almost

impossible to get it to burn. All organic matter contains carbon and hydrogen, and nearly all, oxygen in addition, and all vegetable matter consists of carbon, hydrogen, and oxygen in various proportions, and having various chemical combinations according to the nature and structure of the plant, a few impurities, such as minerals, being usually present. Herbaceous matter contains more than 50 per cent. of carbon, about 5 per cent. of hydrogen, and more than 40 per cent. of oxygen. As it undergoes changes and decomposition, the oxygen decreases, whilst the carbon increases, and the more it progresses towards being formed into coal, so the oxygen decreases and the carbon increases, the hydrogen remaining fairly constant, except when bituminous coal changes into anthracite, and then it decreases.

The following table gives a rough idea of the composition of the coal-forming matter in four stages :

TABULATION 7.

Element.	Wood.	Peat.	Bituminous.	Anthracite.
Carbon	52	60	75	93
Hydrogen	5	5	5	4
Oxygen	43	35	20	3

So that the best coal is composed of nearly all carbon.

1lb. of hydrogen completely burnt, evolves 62,032 heat units.
1lb. of carbon   „   „   „   14,500   „

One heat unit (British) is the amount of heat required to aise 1lb. of water from 60deg. F. to 61deg. F. = 1deg. F., and in mechanical energy it is equivalent to 772·5 foot-pounds. 1lb. of coal gives out about the same heat energy as 1lb. of carbon, hence 1lb. of coal when burnt will produce 772·5 × 14,500 = 11,201,250 foot-pounds of energy.

Let 1lb. of good anthracite coal, when burnt thoroughly, give out heat energy equal to, say, 11,000,000 foot-pounds—that is to say, an amount of energy sufficient to raise a weight of 11,000,000lb. a vertical distance of 1ft. One horse-power signifies 33,000 foot-pounds of work done in the space of one minute of time. If the coal be burnt in the space of one hour, this gives $\dfrac{11,000,000}{60}$ foot-pounds per minute, and dividing this by 33,000, will give $\dfrac{11,000,000}{60 \times 33000} =$ 5·5 h.p., so that burning coal at the rate of 1lb. per hour gives out heat energy equivalent to above 5 h.p., this being the total or theoretical energy given out. A great amount of this energy is wasted, and, roughly, only 50 per cent., or one-half of it, can be said to be utilised. This useful half is employed to evaporate the water in a boiler, and the number of pounds of water that are converted into steam determines what is called the "evaporative power" of the coal, the steam being at an atmospheric pressure and at a temperature of 100deg. C. If the whole of the heat energy of 1lb. of coal were turned to account, it would evaporate about 15lbs. of water; but assuming 50 per cent. of this is lost, only 7·5lbs. of water can be turned into steam, so that the efficiency of an ordinary boiler may be put down at 50 per cent.

The heat given out by burning fuel is called the "total or theoretical heat of combustion," and that portion which is utilised is called the "available or practical heat of combustion." The amount of heat evolved by coal depends a great deal on its quality and chemical composition. Carbon and hydrogen are the valuable elements in fuel, and the percentage of these vary considerably. It need scarcely be said that the choice of coal for a boiler is most important and weighty, and becomes more so as the size and number of boilers increase. The coal bill is always the heaviest item in the working expenses of a Central Station, so it is natural that every buyer should try to obtain the coal that will give a maximum amount of heat for a minimum cost, taking care that the coal chosen is of a nature to suit the particular type of boiler employed. The following table gives reliable and most careful analyses of the chief kinds of coal in Great Britain (compiled by Prof. W. Foster).

TABULATION 8.

PERCENTAGE OF COMPOSITION.

Description of fuel.	Hydrogen.	Carbon.	Moisture.	Ash.	Heat units per lb. of fuel.
Welsh anthracite.	3·5	89·17	1·78	2·12	15,017
Durham coal, caking	5·3	84·34	1·14	2·42	15,221
Yorkshire coal ...	4·93	84·1	2 2	1·2	14,875
English cannel coal	5·02	81·46	2·07	5·8	14,571
Scotch cannel coal	6·18	75·42	4·0	2·24	15,750

Small quantities of nitrogen, sulphur, and oxygen are also present.

Work and Horse-power.

It may not be out of place to make a few remarks and give some practical examples respecting Energy, Force, Power, and Work. These four terms are often used synonymously or applied in a loose manner.

Force may be defined as "that which moves or tends to move matter." When a mass of matter is lying at rest it requires some force to start it, or to overcome its resistance to motion; this resistance to motion is called the inertia of the mass. The weight of a mass of matter is merely the force of gravity acting on it, so if you hold a stone in your hand its weight signifies that the gravitating force of the earth is pulling at the stone, striving to drag it down to the centre of the earth, and the muscular force you have to exert is spent in balancing the force of gravity. The greater the mass is, the greater is the force of gravitation acting on it, consequently its weight is greater. If there were no such thing as gravity, matter would have no weight, and, owing to the enormous centrifugal force due to the revolution of the earth, everything on its surface would tend to fly off into space.

Every particle or mass of matter exerts an attractive force upon every other particle or mass of matter, so that two bodies have a mutual attraction. The amount of attraction a body possesses is in proportion to its mass, and the mutual attraction of two bodies is in inverse proportion to the square of the distance between them—that is to say, two bodies 6ins. apart are attracted to each other by four times the force by which they are attracted when at 12ins. apart. We have every reason to believe that this law holds true in our planetary system, and throughout interstellar space. It is due probably to this theory that the earth is kept at a definite distance from the sun when revolving round it, and that all the planets have their fixed course. So when a mass of lead falls to the earth it signifies that the lead and the earth, following the law of mutual attraction, attract each other with a force proportional to their size, and inversely proportional to the square of the distance between them ; but since the mass of the lead weight is infinitesimal as compared with that of the earth, therefore it follows that the force the lead exerts is comparatively nothing to the force the earth exerts and the result is the lead travels all the distance, and the earth so little that it cannot be perceived.

Fundamental Units.—The French system of measuring quantities is the most simple and beautiful invention that can be imagined. It is greatly to be regretted that this method of measurement is not adopted in England by practical engineers, and that it is not used in workshop practice here. In no branch of science can it be appreciated so much as in that relating to electricity and magnetism. The British workmen still plod on, using the clumsy and unscientific foot and pound, whose use causes a vast amount of brain work and valuable time to be wasted, when both could be saved were they to think and make calculations in the French units. Taking the one example of length. The foot is the British unit of length for engineers, other dimensions are the mile, yard, etc. To express miles in inches, your practical man has to multiply out on a piece of paper the expression $1,760 \times 3 \times 12$. Now examine the French unit of length—namely, the "metre"—this corresponds to the English yard, its equivalent being 39·39ins., or nearly $\frac{1}{10}$th more than the yard. For great lengths, the French use the "kilometre," or 1,000 metres, and so this may be said to correspond with the mile, whilst for short measurements they have the "centimetre," or $\frac{1}{100}$th part of a metre, and this corresponds to the inch, there being 2·54 centimetres in 1in., or one centimetre = $\frac{2}{5}$in. From this it is seen that all dimensions of length are expressed in multiples of the figure 10. Thus a kilometre is reduced to centimetres by multiplying by 100,000, because there are 1,000 metres in the kilometre, and 100 centimetres in one metre. 100,000 can, however, be written in a much simpler and quicker manner by expressing the number in powers of 10; thus 10 can be written 10^1, because there is only one 10, and 100 can be written 10^2, because it signifies two tens multiplied together, or 10×10, similarly $1,000 = 10^3$, and $100,000 = 10^5$, so that one kilometre can be expressed in centimetres by the simple expression 10^5. It is the same with the units of weight. The British units are tons, hundred-weights, stones, quarters, pounds, ounces, pennyweights, grains—an array sufficient to drive anyone to despair. Who does not remember the mental torture of committing to memory those weights and measures tables of our juvenile school days? To express tons in ounces, we must multiply out the lengthy expression of $20 \times 112 \times 16$. The French unit of weight is the "gramme," and $454\frac{1}{2}$ grammes go to the English pound (avoirdupois). The kilogramme signifies 1,000 grammes, and corresponds to the pound, its value being 2·2lbs. The milligramme signifies $\frac{1}{1000}$th part of a gramme, and all these weights are expressed in multiples of 10, same as with the various lengths. One kilogramme $= 1,000,000$ milligrammes, or 10^6 milligrammes. The above instances are only two out of a number that could be quoted to prove how barbarous and ridiculous our units of measurement are. Knowledge is quite hard enough to acquire without making its acquirements doubly laborious; units and standards are merely arbitrary and used as vehicles to convey our thoughts in drawing comparisons, and they should be the slaves of man instead of man being their slave, as he often is.

Length, Weight, and Time are the three fundamental quantities, and in the French system the centimetre is taken as the unit of length, the gramme as the unit of weight, and the second as the unit of time—the symbol letters C.G.S. being used to denote respectively the centi metre, the gramme, and the second. Hence the system is known universally amongst scientists as the C.G.S. system of absolute units. Unfortunately, this system is so little known in England that it is necessary to express mechanical calculations in foot-pounds, since these units are known to everyone, whilst the others are only known to a few.

The Unit of Force is named the "dyne," and it is that

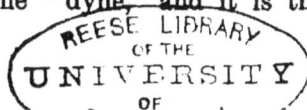

amount of force which acting for one second, when applied to a mass of one gramme weight, will move it through a distance of one centimetre in one second of time. This force is expended in overcoming the resisting force of the mass through a distance, and when this is effected, it is said that *work is done* upon the mass. When a mass is lifted up vertically, work is done upon or absorbed by the mass, since the resisting force of gravity is overcome through a certain distance.

The Unit of Work is named the "erg," and it signifies the work done in overcoming a resisting force of one dyne through a distance of one centimetre. Now if a mass of matter weighing one gramme be allowed to fall freely, it would be seen that, owing to the force of gravity acting upon it, it would fall a distance of 981 centimetres in one second of time; hence the force of gravity equals 981 dynes, so when one gramme of matter is lifted vertically through a distance of one centimetre it is evident that a force of 981 dynes must be overcome through a distance of one centimetre; consequently, 981 units of work must be spent on the mass to raise it that distance.

Energy is an expression that denotes the capability of a body to do work. All moving bodies possess energy, usually termed stored energy; thus, a moving railway train has a vast amount of stored energy in it, and so has a revolving wheel. It has been shown that an expenditure of 981 ergs is necessary to raise one gramme through a vertical distance of one centimetre. It may be thought that this amount of energy is wasted, but this is not so, for we have the law of the conservation of energy to deal with, and the energy is absorbed by the weight and can be recovered; for let the weight fall back to its original position, in the act of falling it will develop energy, and do work, and the work done by it will be exactly equal to the work done upon it; so that a body can have two kinds of stored energy in it, one kind when in motion and the other kind when at rest. All bodies at rest possess "potential energy," in virtue of the position they occupy, as compared with bodies at a lower level. A boulder on the top of a lofty mountain possesses a great amount of potential energy because falling to the plains below it could develop, by virtue of its elevated position, great energy. A boulder on the plains possesses an amount of potential energy measurable by the distance it can fall. From this it can be understood that the potential energy of a body decreases as it gets nearer to the centre of the earth, and if it were possible for a body to be situated exactly in the centre its potential energy would be zero. Kinetic energy, on the other hand, is the energy that is developed by the moving or falling body. A stone just at the moment of falling possesses nothing but potential energy, so that its potential energy is a maximum and its kinetic energy a minimum, or zero ; whilst falling, the potential energy changes into kinetic, half-way down its energy is half potential and half kinetic, and at the moment of striking the ground its kinetic energy, or developed capability of doing work, is a maximum, while its potential energy is at zero, the potential thus becoming converted entirely into kinetic (so far as concerns the potential energy due to the difference between the two levels).

Work can be expressed as the product of two factors, weight and distance ; and however the values of the factors change, provided the product is the same, the work done is the same. Thus, 50 grammes raised 100 centimetres requires as much work as 100 grammes raised 50 centimetres, or 1,000 grammes raised five centimetres.

Power means the rate of doing work or the amount of work that can be done in a certain time.

The unit of power is named the "watt," and signifies

work done at the rate of 10,000,000 ergs per second ; this is
the practical unit, because the absolute unit of power, which
signifies one erg per second, is far too small to be of any
practical value.   It was shown that it required 981 ergs to
raise.one gramme a distance of one centimetre, and as one
watt equals 10,000,000, or 10^7 ergs per second, therefore
(using 1,000 for 981 for simplicity) it will raise $\dfrac{10^7}{10^3} = 10^4$
grammes a distance of one centimetre in one second, or 10
kilogrammes a distance of one centimetre.

One kilogramme equals 2·2 English pounds.
One centimetre   ,,   ·4ins. (nearly).

Therefore 10 kilogrammes equals 22lbs., and 22lbs. raised
.4in. is equivalent to nearly ¾lb. raised 1ft. high.   So in
English units one watt signifies work done at the rate of
lifting ¾lb. a distance of 1ft., or ·75 foot-pound in one
second (approximately).

The watt is adopted as the unit of electrical power, and
will be further considered in this respect when treating of
electrical units.   For mechanical powers the unit of power
in England is the horse-power ; and expressed in units of
pounds, feet, and minutes, it is equivalent to 33,000lbs.
raised 1ft. in a space of one minute—that is, 33,000 foot-
pounds per minute ; and this number was fixed upon because
it was reckoned that a very strong dray-horse could do that
work, but if it could, it could only keep up this rate of work
for a very short time, say, a few minutes, so that the
average power of a good horse, working all day, is far less
than the standard given—in all probability not much more
than 20,000 foot-pounds per minute.   30,000 foot-pounds
per minute = 550 foot-pounds per second, but one watt
equals ·737 foot-pound per second exactly, therefore
550 ÷ ·737 = 746, so that 746 watts go to the horse-power.

There are three different kinds of horse-power used when expressing the power of a steam-engine : Nominal, Indicated, and Brake. Concerning the first—namely, that of nominal—the less said the better ; the word is simply used by makers of steam engines and boilers to express size, and gives no information respecting the power of a machine. Some makers say that their indicated can be taken to signify $2\frac{1}{2}$ times the nominal, others three times ; others, again, twice ; others still something else. It is a wretched term, absolutely useless, and a cause of considerable annoyance ; an undefined term like this always has a suspicious appearance. The disease, unfortunately, has seized hold of electrical apparatus ; thus we have arc lamps of *nominal* candle-power, and the trusting purchaser either finds out, or does not find out (according to his knowledge or ignorance) that the actual candle-power is probably one-half of the nominal. The word " nominal," when applied in this misleading way is deceptive in the sight of every right-minded engineer.

Indicated horse-power signifies the rate of work done by the piston, and so is a measure of the power developed by the engine. Brake horse-power signifies the power that is given right off the driving wheel, and evidently this measure is the best and fairest, because it represents the actual useful power given by the engine. The brake is necessarily less than the indicated, since the difference between the two measures the loss due to the friction, etc., of the moving parts, and this loss differs a great deal in various engines, according to whether they are well designed or not ; it also differs according to the size of the engine. For example, a large-sized engine may have its brake 12 per cent. lower than its indicated, while in small engines it may be 20 per cent. or more.

· *How to Calculate Indicated Horse-power.*— This is done in the following way : Multiply the average pressure of steam in pounds per square inch by the area of the piston in square inches ; this gives the total force acting on the piston. Now multiply this by the distance in feet through which the piston moves, or the length of stroke, as it is termed, but since the piston must make a forward and backward stroke in order to produce one complete revolution of the driving wheel, hence the length in feet must be multiplied by 2. We have now obtained the product of a force acting through a distance, which signifies work done in foot-pounds ; and to obtain the power, or rate of doing work, we must finally multiply by the number of revolutions per minute that the driving wheel makes. This gives the number of foot-pounds moved in the space of one minute, and when the rate of doing work is such that 33,000 foot-pounds of work are done in one minute, then 1 h.p. is produced, so we have now to divide the total product by 33,000. The result gives the horse-power.

This rule may be expressed in the following formula :

$$\frac{P\,L\,A\,N}{33,000} = 1 \text{ h.p.}$$

P = pressure of steam in pounds per square inch.
L = length of stroke of the piston in feet, multiplied by 2.
A = area of the piston in square inches.
N = number of revolutions per minute of driving wheel.

Here the three fundamental units of weight, length, and time are represented.

1. Pressure representing the weight in pounds,
2. Stroke　　　　,,　　　,, length in feet.
3. Revolutions ,,　　　,, time in minutes.

The following example shows how to apply the above : The cylinder of a simple engine is 9ins. in diameter, its length 12ins., speed is 200 revolutions per minute, and average steam pressure 80lbs. per square inch. Find its indicated horse-power.

Since the cylinder encloses the piston, the diameter of it (internally) is that of the piston.

$$\text{Area} = \frac{\pi \, d^2}{4}, \text{ where } d = \text{diameter,}$$

$$\text{and } \pi = 3\cdot1416.$$

$$\therefore \text{Area} = \frac{3\cdot1416 \times 81}{4} = 63\cdot61 \text{ square inches.}$$

Let the length of the stroke be 11ins., then for the forward and backward stroke the total length will be $11 \times 2 = 22$ins. $= 1\cdot83$ft. ; speed is 200 revolutions per minute ; average pressure is 80lbs. per square inch. Thus we have

$$P = 80, \ L = 1\cdot83, \ A = 63\cdot61, \ N = 200,$$

or, $$\frac{80 \times 1\cdot83 \times 63\cdot61 \times 200}{33,000} = \text{I.H.P.} = 56\cdot4.$$

Hence the engine will yield 56·4 I.H.P.

When the engine is compounded the steam is admitted into the small high-pressure cylinder first, and the exhaust steam, which now has a very much lower pressure, then goes into the low-pressure, or large cylinder, so that two calculations must be made and added together—one for the work done in the high-pressure cylinder, and the other for the work done in the low-pressure cylinder.

The following tabulation gives values of Force, Work, and Power expressed in English and French Units of Weight and Length.

TABULATION 9.

FORCE.

1 Milligramme weight represents (nearly)	......	1 dyne.
1 Gramme	,,	,,	......	981 dynes.
1 Pound	,,	,,	 444,980	,,

E

WORK.

1 Milligramme-centimetre does work = (nearly)		1 erg.
1 Gramme-centimetre ,, ,, =		981 ergs
1 Foot-pound ,, ,, =		13,562,859 ,,

POWER.

10^7 Ergs per second .... =		1 watt.
1 Kilogrammetre per second ...... =	(nearly)	10 watts.
1 Foot-pound ,, ,, ... =		1·35 ,,
75 Kilogrammetres ,, ,, ... = (French H.P.)		736 ,,
550 Foot-pounds ,, ,, ... = (English H.P.)		746 ,,

Mechanical Equivalent of Heat.

Heat is a form of energy, and no work can be done without the production of heat. A revolving wheel contains kinetic or stored up energy, and if suddenly stopped by applying a brake, heat is at once produced, due to the tremendous friction that exists. It was of the utmost importance to know the relation that exists between the unit of heat and the unit of mechanical energy. This want was supplied by Joule, who conducted most elaborate experiments for six years in order to obtain accurate results. The chief method used by Joule was concerning the friction of fluids, and he obtained his data in the following way : A number of paddles were delicately suspended in a vessel of water, a length of string was wound round the spindle of the paddles, passed over a pulley, and then attached to a weight. Upon allowing this weight to fall, rapid motion was given to the paddles, and the great agitation of the water raised its temperature. The mechanical power was obtained by multiplying the known weight by the distance it fell, and the equivalent amount of heat created was measured by the rise of temperature of the known mass of water.

These experiments established without doubt the following

law connecting heat and work : *The amount of energy produced as heat is an exact measure of or is equal to the amount of energy expended mechanically.*

Joule found that the work expended by allowing a mass of 424 grammes to fall by gravity a distance of one metre would produce sufficient heat to raise the temperature of one gramme of water through one degree centigrade (say, from 9deg. C. to 10deg. C.). This amount of heat is fixed upon as the unit of heat, and is called a Calorie. Since one gramme falling one centimetre = 981 ergs of work, therefore 424 grammes falling one metre = $981 \times 424 \times 10^2$, or 41,594,400 ergs, so that one Calorie of heat may be taken to be the equivalent of 42,000,000 ergs of work done. It was shown that one watt, the practical unit of power, was equal to 10,000,000 ergs of work done per second; therefore power, or rate of doing work, can be expressed in heat units, and one calorie per second $= \dfrac{42,000,000}{10,000,000} = 4 \cdot 2$ watts,

or, 1 Watt $= \cdot 24$ Calorie per second.

In electrical work this last formula is a very important one, for by its use, as will be seen later on, it enables the electrician to calculate the extent to which electric mains and branch wires, and also the coils of dynamos, motors, etc., are heated, when a given amount of watts of electrical power is passing through them. It is a formula that should be constantly in the mind, because the production of electricity and its distribution, whether in the form of Light or Power, is invariably accompanied by Heat. Electricity, Heat, and Light, all three, are closely identical in their phenomena. From electricity, heat and light are produced; from heat, electricity and light are produced; from light, electricity and heat are produced. Magnetism also exhibits phenomena that lead us to identify it to some degree with the above three, because from magnetism electricity can be produced, and also heat.

As an example of this latter statement, take the case of a bar of iron, which will become rapidly heated by magnetising it by means of an alternating current of electricity. With respect to the relations between magnetism and light, we have Maxwell's " Electromagnetic Theory of Light." This theory arose from the belief of electricity and magnetism being produced by motion of the ether (the ether being the name given to that which is supposed to pervade all bodies and all space), the same with regard to the propagation of light-rays. An extraordinary fact is that the velocities of propagation of electromagnetic induction and of light are so close to each other as to be almost identical, the velocity of electromagnetic induction being $2 \cdot 9857 \times 10^{10}$ centimetres per second, and the velocity of light being $2 \cdot 9992 \times 10^{10}$ centimetres per second. A practical reason for this identity is that whenever any disturbances of the sun's surface take place, magnetic disturbances immediately take place on the earth's surface, and affect the working of telegraphic apparatus.

It is useful to know the mechanical equivalent of heat when expressed in English units. The English unit of heat is that amount of heat which will raise 1lb. (pound av.) of water from 60deg. F. to 61deg. F., or 1deg. F., and the mechanical energy required to produce one unit of heat is the work done by a weight of 1lb. falling by gravity through a distance of $772 \cdot 55$ft., or, in other words, $772 \cdot 55$ foot-pounds.

Hence $772 \cdot 55$ foot-pounds = one unit of heat. This amount of mechanical energy is known as " Joule's equivalent," or by the letter " **J**."

Plotting Curves.

The method of showing the relations existing between two variable quantities by means of a curve traced on squared paper is one that cannot be too highly commended, and it is one that is not used so much as it should be, since, being essentially practical, it appeals at once to the engineer.

In no branch of engineering can curve tracing be shown to greater advantage than in electrical work. It is invaluable in showing at a glance the laws and actions of all manner of apparatus, efficiencies of dynamos, lamps, etc., lighting loads of central stations, charging and discharging accumulators, magnetic effects, and a multitude of things too numerous to mention. By means of curves we can see and grasp in a moment all that we know about the particular object under examination, and in a manner that cannot be presented by mere columns of figures.

In plotting curves, it is necessary to use what is called "squared paper"—this paper is divided up into small squares by a number of parallel faint lines running across the paper, both vertically and horizontally. The paper can be obtained with squares of various sizes, in fractions of an inch or centimetre, according to choice, the size mostly used being a square having sides about $\frac{1}{4}$cm., or $\frac{1}{10}$in. These small squares are contained in larger squares having sides usually five times the size of the small squares. These large squares are mostly denoted by thick blue and red lines, so one large square embraces 25 small ones, the red lines running at right angles to the blue. When the curve to be plotted is a long one, or only requires to be done roughly, then the large squares are used with advantage, but where the curve is short, or requires to be done very carefully by plotting small increments, then the small squares must be used. The horizontal base line is named the axis of x, since the various values of x are marked off along this line. The vertical line is named the axis of y, since the various values of y are marked off along this line. Both axes start from one point, named the origin. When the axis of x is extended to the left of the vertical axis, then values of x plotted on that side of zero are considered as of negative value, and so have the minus sign given them, thus : $- x$; similarly, when the axis of y is extended to below the horizontal axis, then values of y plotted on that side of zero are considered as of negative value, also have the minus sign given them, thus : $- y$; so that a line is considered positive or negative according to its direction from the origin. Here we will only consider the positive values of x and y, or those plotted to the right of zero and above zero respectively. Having mapped out a suitable scale along x and y, the values of x and y are now plotted. This is done as follows : The first value of x is taken, and a mark made on the horizontal axis, where the scale gives the same value or the nearest value that can be judged. The first value of y is now taken, and similarly noted on the vertical axis. The position of a point anywhere within the space enclosed by the two right lines, x and y, will give a value for x and a value for y. This point is found in this way : first project the mark on the horizontal line in a vertical direction, measured by the value for y ; then project the mark on the vertical line in a horizontal direction, measured by the value for x. The position required will be that given by the inter-ception of the two lines, drawn at right angles from x and y, from the defined marks on the horizontal and vertical axes respectively. These two lines are shown dotted in Fig. 2. The point, p, of their interception gives the plotted position of values for x and y. The next two values are now taken and positioned, and so on with all the rest ; the positions in practice being denoted by a small distinct cross.

When the ratio existing between corresponding values for x and y is constant throughout, then all the points of the curve will lie in one straight line, and therefore can be joined together by drawing a line through them. The angle which this straight line makes with the horizontal axis is termed the "slope" of the curve, and it varies as the ratio between x and y varies. In cases where the rate of increase of y is small as compared with the rate of increase of x, then the slope will be horizontally inclined—that is, it will form some angle between 45deg. and the horizontal; on the other hand, where the rate of increase of y is great as compared with that of x, then the slope of the curve will be

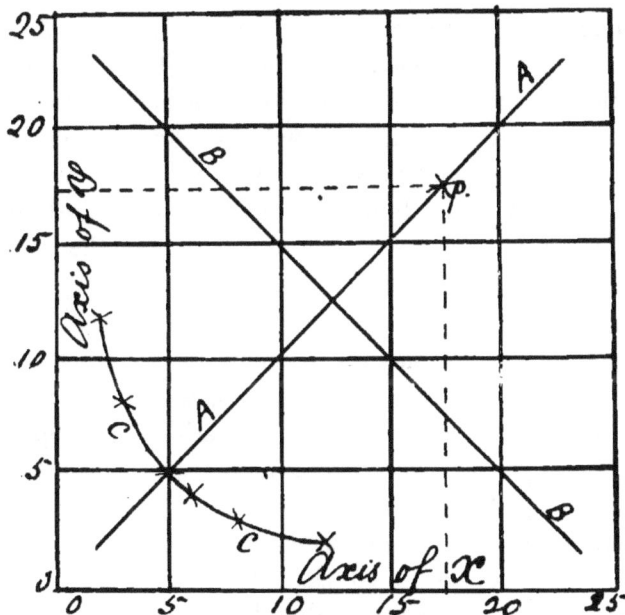

FIG. 2.

vertically inclined, or between 45deg. and the vertical; lastly, where the variables, x and y, both increase at the same rate, then the slope will be midway, or at an angle of 45deg.

The relation that exists between the two variables x and y, is termed the "law" of the curve; and where this curve is a straight line, as just mentioned, the law can be expressed by the very simple equation $\frac{y}{x} = k$, where the fraction $\frac{y}{x}$ signifies the ratio between the two variables, and k signifies a constant. But the fraction $\frac{y}{x}$ is the tangent of the angle (a) which the straight line makes with the horizontal.

$$\therefore \frac{y}{x} = k = \tan a,$$

and $y = x \tan a$, where a signifies the angle of slope.

In Fig. 2 the line A A has a slope of 45deg.; and since tan 45deg. = 1, $\therefore y = x \times 1 = x$; thus proving that x and y both increase at the same rate. The curve B B is also straight, with the same slope as A A, but in a negative direction, its angle being − 45deg. Here it will be noticed that the sum of any two corresponding values of x and y is always the same—*i.e.*, it is a constant—hence $x + y = k$; and the law to the curve is $y = k - x$, where k signifies the constant, which in this particular case is 25.

In dealing with curves other than straight, the laws connecting the two variables are much more complicated, and, in fact, some are so erratic in their courses that it is almost impossible to find their law, or equation, although it is always possible to give some general law which approximately or partially solves the problem. A simple example of a curve is that given when the product of the two variables is a constant; so that when any value for x is doubled, the corresponding value for y is halved, and *vice versâ*. The result of plotting two variables, following this law, is shown by the curve C C in Fig. 2; here it is seen that when $x = 2$, $y = 12$, and when $y = 12$, $x = 2$, and so on, the values for x increasing at the same rate as the values for y decrease, putting $xy = $ constant $= k$, $\therefore y = \frac{k}{x}$, and inserting the value of k we have $y = \dfrac{24}{x}$, which is the law for the curve C C. There are several most important natural laws in physics which may serve as examples of this class of curves, notably Boyle's law connecting the pressure and volume of a gas at constant temperature ; in this instance, as is well known, decreasing the pressure increases the volume, and *vice versâ*, and the product, pressure × volume, is constant. Again, there is the example of a horse-power, whose components are feet and pounds; or, in electrical matters, there is the kilowatt; here, amperes can be plotted along the axis of x, and volts along the axis of y, and the product of any two corresponding values for x and y will give a kilowatt.

Sufficient has now been said respecting curve plotting, and it can readily be understood that all manner of curves can be obtained by plotting the values of two varying quantities.

Steam Boilers, Engines, Etc.

In choosing a type of boiler, two of the most important things to go by are :

First. What sort of water will be used.

Second. What sort of fuel will be used.

All water has got impurities in it to a greater or less extent. The most common and frequent impurities met with are sulphate of lime, carbonate of lime, silica, chloride of sodium, oxide of iron, etc.; also organic matter. These impurities are deposited on the inside of the boiler shell and so form a dangerous deposit, called boiler scale. This scale is of a soft or floury nature when deposit is largely made by carbonate of lime, and is of a hard or stony nature when made chiefly by sulphate of lime.

Where the water is free from impurities, then a locomotive or multitubular boiler may be used ; but where the nature o the water is such as to contain considerable quantity of scale-forming impurities, then only a Lancashire type should be used, because these can be easily cleaned and inspected. With regard to the coal likely to be used, where it is very expensive, then a multitubular boiler may be used with advantage on account of the greater economy of fuel, and where coal is plentiful a Lancashire could be used. But, in addition to water and fuel, there are a number of other considerations that must be carefully weighed in selecting a boiler. Amongst them, there is the nature of the work to be done, long running hours or short running hours, amount of space available, whether locality be on a country estate or in centre of a town, whether skilled attendance is available or not. All these conditions go towards determining the kind of boiler that would be most suitable. It is impossible to lay down any rules to be followed in the choice of a boiler; experience and practice must decide that. The classification which follows gives a few guiding notes in a succinct form, but subject to broad interpretation. The twelve columns give the most important points. Those in row A signify that a Lancashire type of boiler is preferable, and those that are placed in row B signify that a multitubular boiler is preferable. Where the conditions are partly in A and partly in B, as it will be in most cases, then the advantages of the one must be weighed against those of the other.

An average fuel consumption for a stationary locomotive boiler is about 20lbs. of good steam coal per hour for every square foot of grate area, and the water evaporated is about 10lbs. per pound of coal.

In selecting the size of boiler to supply steam to an engine it is necessary to know the quantity of steam required per horse-power, for the capacity of a boiler for doing work is usually expressed in the number of pounds of water that can be evaporated per hour.

For non-condensing engines working at 70lbs. pressure, the standard mostly adopted is that 30lbs. of water should be evaporated per hour for 1 h.p.

TABULATION 10.

	Type.	Water.	Fuel.	Floor rent	Locality.	Attendance.	Cost of repairs
		1	2	3	4	5	6
A.	Lancashire with Galloway tubes.	impure	cheap	cheap	Country or remote places.	intelligent	slight
B.	Multitubular.	pure	expensive	expensive	in towns.	skilled	heavy

	Type.	Efficiency	Prime cost.	Setting.	Space per unit capacity.	Getting up steam.	Life.
		7	8	9	10	11	12
A.	Lancashire with Galloway tubes.	good	cheap	expensive	large	slow	long
B.	Multitubular.	better	expensive	cheap	small	quick	short

For good economical engines, Prof. Thurston gives the following rule for rough approximation : Divide the constant 200 by the square root of the pressure per square inch ; the quotient gives the number of pounds of water that must be evaporated per hour to produce 1 h.p. This rule is written thus—

$$\frac{200}{\sqrt{p}} = W ;$$

where p = the steam pressure per square inch, and W = the weight of water in pounds.

When the very best engines are used, the constant 150 can be used instead of 200.

Suppose a steam-engine of 80 h.p. is to be run at a pressure of 100lbs. per square inch, and it is required to know the size of boiler ; using the constant 200, we have

$$\frac{200}{\sqrt{100}} = W = \frac{200}{10} = 20 ;$$

So that this engine will require 20lbs. of water to be evaporated for each horse-power ;

and the total quantity required = 20 × 80 = 1,600lbs.

So the boiler must be of such dimensions that it can evaporate 1,600lbs. of water per hour, at a pressure of 100lbs. per square inch. From this it is seen that the higher the pressure at which an engine works, the less is the quantity of water required per horse-power.

Steam power is by far the most reliable motive power, particularly for electric lighting purposes, where the greatest steadiness in running is required.

Steam-engines may be either simple or compound, condensing or non-condensing.

In *Compound Engines* the steam, after doing its work in the cylinder, is led into a second cylinder at a lower pressure. The first is called the high-pressure cylinder, and the second the low-pressure cylinder. There are also triple-expansion engines, when the steam passes successively through three cylinders, and lately quadruple-expansion engines are being built, where four cylinders are used ; but this class is only used for marine purposes, and the triple-expansion is very little used on shore, it being the general marine type. A simple engine uses chiefly low-pressure steam, say from 50lbs. to 80lbs. per square inch, and the more the steam is used expansively, the higher, as a rule, is the pressure used ; so that a compound engine will use steam from 80lbs. to 120lbs. per square inch, and a triple-expansion engine will use from 120lbs. to 160lbs. per square inch.

An ordinary compound engine will save 15 to 20 per cent. of fuel, and a triple-expansion engine will save 15 to 20 per cent. of the fuel used in an ordinary compound, or about 30 per cent. altogether. Where small power is used, then a simple engine will suffice, and as the power required becomes larger, so compound and triple-expansion engines can be used with advantage.

Non-condensing Engines.—This name is applied to all those engines, whether simple or compound, that send the exhaust steam into the atmosphere, which is evidently a great waste of heat.

Condensing Engines.—In 'these the exhaust steam is directed into a condenser, which is a large iron tank, around which a constant flow of cold water is maintained; upon entering, the steam is at once condensed into water, due to the low temperature of the tank.

Running Cost of Steam-Engines.

Using 4lbs. of coal to the brake horse-power per hour, with coal at 20s. per ton, the cost of 1 b.h.p. per hour works out to ·43 or $\frac{3}{7}$ penny; at 15s. per ton, to $\frac{1}{3}$ penny; and at 12s. per ton, to $\frac{1}{4}$ penny. This is for coal alone, which varies in price according to the locality. Other expenses, as oil, wages, water, etc., can be taken as fairly constant in price. It is very difficult to predict the annual cost of motive power; there are so many varying items which depend on local circumstances. An approximate idea can be given, however, which may be found useful. There are also great differences existing in the working expenses of engines of various sizes which are altogether out of proportion, and another great controlling feature is the number of hours per annum run; here, again, the expenses are out of proportion. A third very important point relates to the load an engine has; the more an engine works towards its full load the less is its proportional cost. An example will now be given of the probable annual running cost of a steam plant giving 40 b.h.p., working at full load, and running 10 hours per day, or about 3,000 hours per annum. The total cost can be divided into two equal parts, one of which will represent depreciation, interest, repairs, and wages. This half may be regarded as a fixed amount, independent of what the running hours may be. It may be argued that only wages and interest should be held as of fixed value, since the wear and tear will vary the depreciation and repairs. Strictly speaking, this is correct, but, to simplify matters, it is assumed that they remain constant. The other half of the total cost is put down to coal, oil, waste, water, and sundries—the cost of an engine and boiler, including a fair estimate for the expenses of properly fixing same, being taken at £15 per brake horse-power, or £600 for 40 b.h.p.

TABULATION 11.

A.—INVARIABLE COSTS.

	B.H.P. per annum.	B.H.P. per hour.
Interest on outlay, £600 at 4 % ...	£0 12 0	·048 penny.
Depreciation ,, ,, 7½ % ...	1 3 0	·092 ,,
Repairs ,, ,, 5 % ...	0 15 0	·060 ,,
Wages of driver, £100 per annum	2 10 0	·200 ,,
Total	£5 0 0	·400 penny.

B.—VARIABLE COSTS.

4lbs. of coal per hour for 3,000 hours, using coal at 15s. per ton	£4 0 0	·32 penny.
Water, oil, waste, and sundries ...	1 0 0	·08 ,,
Total	£5 0 0	·40 penny.

So that total cost per brake horse-power per annum = £10.

and ,, ,, ,, ,, hour = ·8 penny

If the above plant be used for working the electric light, the running hours will be very much shorter. Take an average time, say from dusk until 10 p.m., this will be about 1,500 hours throughout the year, including Sundays. This is one-half of the preceding case, but the cost will not be one-half, because part of the total cost remains constant, and this amount is £5 per brake horse-power per annum. The other part of the cost will now be in proportion to the hours run, so we have $\dfrac{5 \times 1,500}{3,000} = 2\frac{1}{2}$. Thus the total cost for 1,500 hours = £7. 10s. per brake horse-power per annum, or 1·2 pennies per brake horse-power hour. If the hours were shorter the proportional cost would be greater, and if longer it would be less. When the cost per hour works out very small, then the plant is said to possess great earning capacity, and this applies to all machinery, whether steam-engines, gas-engines, dynamos, or what not. This point is a very serious question in central electric light stations, for the expensive plant has to be idle all day, and is only earning money on an average of half-a-dozen or so hours per day out of the 24.

We will now investigate the running cost of small-sized steam-engines. The outlay is higher per horse-power in the first place, say £20 per brake horse-power, or £200 for a 10 b.h.p. steam plant; wages remain same. The economy will be less, so that 25 per cent. more for coal, oil, etc., must be allowed than for the larger plant.

Tabulating these results, we get

TABULATION 12.

A.—INVARIABLE COSTS.

	B.H.P. per annum.	B.H.P. per hour.
Interest on outlay, £200 at 4 %	£0 16 0	·064 penny.
Depreciation ,, ,, 7½ %	1 10 0	·120 ,,
Repairs ,, ,, 5 %	1 0 0	·080 ,,
Wages of driver, £100 per annum	10 0 0	·800 ,,
Total	£13 6 0	1·064 pence.

B.—Variable Costs.

5lbs. of coal per hour for 3,000 hours,
    using coal at 15s. per ton............ £5 0 0   .... ·4 penny.
Water, oil, waste, sundries, etc, ... 1 5 0   .... ·1 ,,

        Total.................. £6 5 0   .... ·5 penny.

So that total cost per brake horse-power per annum = £19. 11s.
   and   ,,   ,,   hour = 1·56 pence.

This is just about double the working cost of the large
plant per brake horse-power hour, or looking at it from
another point of view, the large plant will yield four times
the power at only double the cost, since $40 \times £10 = £400$,
and $10 \times £20 = £200$.

Now let this small plant run for only 1,500 hours per
annum, we have £13. 6s., which remains constant, and a
proportional cost for the rest.   That is

$$\frac{6\frac{1}{4} \times 1,500}{3,000} = £3\frac{1}{8} = £3. \text{ 2s. 6d.}$$

and £3. 2s. 6d. + £13. 6s. = £16. 8s. 6d. per annum.
This is equal to 2·6 pence per brake horse-power per hour.

The Gas-Engine.

Those who visited the Electrical Exhibition held at the
Crystal Palace during the first half of 1892, and who
inspected the Machinery Department, must have been
surprised to see the great number of gas-engines that were
at work—gas-engines of all sorts and conditions. There
probably has never been such a display of gas-engines of
various types (more than a dozen) before ; at all events, not
in England.

The advent of the gas-engine could not have occurred at
a more opportune time than in conjunction with the advent
of electric lighting ; the one has to a great extent further-
anced the adoption of the other, and *vice versâ*. The private

·electric light installations, such as in country houses, hotels, ·etc., called for a reliable source of small power which could be drawn upon at five minutes' notice, and did not necessitate constant individual attention. This want was supplied by the gas-engine. On the other hand, it is safe to say that the adoption of electric lighting has been the means of considerably widening the field in which this kind of motive power can be used with advantage. The great number of gas-engines of various types now invented is due to the ·expiry of patent rights in England of the fundamental principle of Otto's cycle, hitherto held by Messrs. Crossley Bros., and which expired in the summer of 1890. The idea of ·obtaining motive power by igniting explosive gases dates back to the eighteenth century, when Robert Street, in 1794, was the first to make a practical gas-engine having a cylinder and a piston. After this, numerous attempts and inventions were made to produce a gas-engine that would be of practical and commercial value. But it was not until the year 1876 that the internal-combustion engine assumed a shape that made it at once a reliable source of motive power and a sound commercial success. This important improvement was made by Dr. Otto, whose name is now famous all over the world. The manufacture of these engines was taken up in England by Messrs. Crossley Bros., of Manchester, who have made great improvements in detail and mechanical construction.

The Otto Cycle.—The method in which Dr. Otto uses the gas, and the operations that take place in the cylinder of the engine, is known as Otto's " cycle of operations," and in order to understand how a gas-engine works, it is necessary to know what the "cycle" is. The cycle of operations consists of four distinct operations. We will imagine the piston at the top end of the cylinder ready to make its forward stroke. Upon moving forward, the admission-valve is

F

opened, and a mixture of coal gas and air is drawn into·
the cylinder after the piston. When the piston reaches the
bottom of the cylinder the admission-valve is closed, the
whole of the cylinder space in front of the piston being now
occupied by a gaseous mixture. Upon the piston making its
return stroke it drives all this mixture in front of it, and, as
it cannot escape, it is compressed, so that by the time the
piston has reached the end of its stroke the gas exists under
great pressure. These two strokes of the piston, one forward
and one return, constitute one complete revolution of the
flywheel. The third operation now takes place—that is, the
ignition-valve is opened, and the mixture fired. The explosive force gives an impetus to the piston, which consequently
travels forward; upon arriving at the end of its stroke, the
impetus given to the flywheel tends to carry it round past
the dead centre, therefore the piston is urged to make the
return stroke. At the moment of returning, the exhaust-
valve is opened, and so the fourth operation consists of the
piston driving out before it all exploded mixture, which now
escapes by the exhaust-port. Upon arriving at the end of
its return stroke the admission-valve is again opened, ready
to admit a fresh charge of gas and air, and so the first operation is ready to be repeated. These four operations constitute a complete cycle, and can be stated thus :

A. Admission of gaseous mixture.
B. Compression   ,,   ,,
C. Expansion   ,,   ,,
D. Exhaustion   ,,   ,,

The cycle produces two complete revolutions, so that an
explosion takes place and an impulse is given to the piston
once every two revolutions, or during every alternate forward
stroke of the piston.

Combustion in the Cylinder.—When a gaseous mixture

explodes, it means that the gases burn with very great rapidity, so that explosion signifies rapid combustion. The intensity or force of an explosion depends on a number of circumstances, such as the composition and nature of the gases, their pressure, etc. The duration of an explosion is the time that elapses between ignition and the maximum pressure produced. The shorter this time is, the more powerful and effective is the explosion. Further, this maximum pressure occurs, and therefore the explosive character ceases, when one-half of the total heat of combustion has been evolved. A rich gaseous mixture explodes more rapidly than a poor mixture. After ignition the whole of the heat of combustion is not evolved at once, but takes place gradually, combustion taking place more slowly the nearer the end is reached; for if this were not so, the shock would be so great that enormous pressure and strain would be put upon the mechanism of the engine. Combustion and evolution of heat thus go on throughout the whole length of the piston stroke. By making the combustion gradual in this manner, a more even and steady pressure is given to the piston. An explosion every revolution gives a more steady-running machine than one in every other revolution.

Efficiencies.—The power and efficiency of a perfect heat-engine can be measured by the absolute temperatures between which it works. When mechanical energy is produced from heat energy, it must always be accompanied by the annihilation or disappearance of an equivalent amount of heat energy, and this is shown by the fall of temperature which takes place in the heat-producing body. The higher and lower temperatures between which the fall occurs are called the "limiting temperatures," and the greater the difference between them the greater is the power produced. From this it is clear that when a body does work it must

F 2

lose heat or fall in temperature, just as a body of water, in order to do work, must fall from a higher to a lower level, or a current of electricity from a higher to a lower potential. The limiting temperatures indicate the "efficiency" of the perfect heat-engine, and the greater the difference of temperatures the greater the efficiency. Take the case of an ideal steam-engine, where the temperature of the steam on entering the cylinder is, say, 200deg. C., and on leaving is 50deg. C. The fall in the temperature measures the work that is done on the piston. First, however, the temperatures must be reduced to absolute temperature. Absolute temperature is −273deg. C.; that is to say, a body at that temperature is supposed to have so little heat that it is fixed theoretically as zero; so that a body at the ordinary temperature of 0deg. C is really at 273deg. C absolutely.

The limiting temperatures absolutely are then (200 + 273) and (50 + 273) or 473 and 323.

Let the higher temperature be T,
,, lower ,, t,

then the efficiency of any heat-engine working between these two limiting temperatures can be calculated by the general formula—

$$\frac{T - t}{T} = \text{efficiency.}$$

Applying this particular example, we have—

$$\frac{473 - 323}{473} = \text{efficiency} = \cdot 31.$$

Hence an engine working between these temperatures can only utilise about one-third of the heat supplied to it, and this is the utmost efficiency, theoretically, that can be got out of it, judging it an ideal or perfect engine.

The steam-engine in use to-day is very far from being perfect, and heat is wasted to a great extent, as the best made can only utilise 20 per cent., and if judged from the coal consumed, the total efficiency is only 10 per cent., since 50 per cent. is lost in the boiler.

In a gas-engine the cylinder is the furnace of the engine, and the fuel and the motive force are one and the same, whereas in the boiler the fuel is coal and the motive force steam.

According to Prof. Robinson, in his authoritative work on "Internal-Combustion Engines," the temperature of the exploding gases in the cylinder of a gas-engine can be taken at about 1,600deg. C. just after the moment of ignition, although it is very difficult to find out what is really going on inside the cylinder, there being so many disturbing elements with which to contend. As the gases burn and expand, the temperature falls, until just at the end of the stroke, when they are leaving the exhaust-port, the temperature is about 400deg. C. This gives an ideal efficiency of

$$\frac{1,873 - 673}{1,873} = \frac{1,200}{1,873} = \cdot64.$$

Like the steam-engine, a great amount of this heat is lost in practice; the exhaust gases carry off nearly one-half of the total heat available, while the water-jacket carries off another quarter. The following proportions are given by Prof. Robinson as to the probable way the heat is distributed:

Heat utilised by piston .... 23 per cent.

„ carried off by exhaust .... 27 „

„ „ „ water-jacket...... 50 „

Total heat of combustion = 100

Comparing the total efficiency of the steam-engine with that of the gas-engine, a steam-engine utilises, say, 10 per cent. of the heat evolved by the coal; a gas-engine utilises, say, 20 per cent. of the heat evolved by the coal gas. So that a gas-engine is twice as efficient as the steam-engine. A very good steam-engine uses, say, 2lb. of coal per brake horse-power hour, this being the utmost efficiency practicable ; 1lb. of coal, when burnt, will evolve 14,500 heat units; 2lb. = 29,000 units. In a large-sized gas-engine it may be taken that the consumption of gas per brake horse-power per hour is 22·5 cubic feet; 30 cubic feet of gas weigh 1lb., and evolves 20,000 heat units ; 22·5 cubic feet of gas weigh ·75lb., and evolves 15,000 heat units. Therefore the gas-engine only requires 15,000 heat units to produce 1 b.h.p., whilst the steam-engine requires 29,000, or close on double.

The efficiency of the mechanism of a gas-engine—that is, the ratio of the indicated horse-power to the brake horse-power—is high, and about similar to that of a steam-engine. On an average it may be put down that the brake horse-power is 85 per cent. of the indicated horse-power, or that 15 per cent. is lost in friction of the moving parts, etc. One pound of good anthracite coal will yield about 4·5 cubic feet of coal gas, so that in fairly large gas-engines the consumption of coal per brake horse-power per hour is, say,

$$\frac{22 \cdot 5}{4 \cdot 5} = 5\text{lb}.$$

The heat energy obtained from burning the gas obtained from 1lb. of coal is therefore $\dfrac{20,000}{30} \times 4 \cdot 5 = 3,000$ heat units, or only one-fifth of the heat energy obtained by burning the coal

The Water-Jacket.—The great temperature attained by the exploding gases, about 1,600deg. C., and the great amount of heat that is absorbed by the walls of the cylinder, would have a most injurious effect on the cylinder, causing overheating, expansion, etc., and preventing lubrication, etc., unless it were protected in some way for this purpose. The cylinder is enclosed in a water-jacket, somewhat like a steam-jacket round the cylinder of a steam-engine; through this water-jacket a current of cold water is caused to circulate. To obtain a continual circulation of water, a large tank is fixed as close as convenient to the engine, and at a level of, say, 6ft. above the floor. A pipe leads from this tank into the bottom of the water-jacket, and another pipe leaves the top of the water-jacket and returns to the water-tank. The water has thus a free passage round the walls of the cylinder, the great heat from which raises the temperature of the circulating water. The circulation is produced by the water leaving the jacket hotter than it enters it. In general, the water leaving the jacket ranges from 40deg. C. to 60deg. C., according to the type of engine, and the water entering should be a little above the temperature of the atmosphere. A suitable tank is always fixed up by the supplier of the engine.

Consumption of Gas.—It may be taken that a medium-sized gas-engine—say 14 b.h.p.—will consume about 25 cubic feet per brake horse-power per hour, working at full load, when the gas is supplied at a good pressure and is of a good quality. In districts where the gas is poor in its heat-giving properties and below the proper pressure, the consumption will be greater. This is often the case in country districts or about small towns. The consumption of gas in a gas-engine may be taken to vary inversely as the power of the engine. Taking a 14 b.h.p. as a medium and standard, we may put down 25 cubic feet per brake horse-power per hour at full load. Now as the power rises so the consumption of gas falls, and as the power falls so the consumption rises. The following tabulation will serve as a guide, although the figures given are only approximate, and may be slightly higher or lower, according to the type of gas-engine used, no particular gas-engine being specified.

TABULATION 13.

Brake horse-power	3	6	9	14	22	35	72
Cubic feet per b.h.p. per hour..	37	30	27	25	22	20	17

It is not advisable to work a gas-engine up to its utmost power unless circumstances demand it; a fair margin should always be allowed. This is particularly so when it is working electric lights, because if an explosion is missed, as often happens, the engine becomes overloaded, and tends to pull up, and very often it must be eased before it can recover itself. Lately, gas-engines of very large power have been put on the market, and with great success, sizes of 200 i.h.p. being manufactured. Engines of above 80 i.h.p. or about that power are made with twin cylinders. The very small sizes, say from 1 b.h.p. to 6 b.h.p., are made both vertical and horizontal, the vertical design being very useful where floor space is limited.

Ignition Tube.—The gaseous mixture is fired in the modern gas-engine by means of an ignition tube, and the following account will explain how it is done : The tube is a short piece of ordinary wrought-iron gas-piping, $\frac{3}{8}$in. to $\frac{1}{2}$in. diameter and about $7\frac{1}{2}$in. long, the top end of the tube being closed. The tube is placed over the flame of a Bunsen burner, and so made red hot, the lining of the chimney covering the tube consisting of asbestos. The top of the chimney is closed by a perforated cover, so as to prevent any dust or foreign matter falling in ; a tap regulates the amount of gas in the Bunsen burner, and in practice this tap must be adjusted so that the tube attains a cherry-red heat, which can easily be seen by glancing down the chimney, and on no account must this degree of heat be altered. Any excess will quickly burn out the tube, which signifies stoppage of the engine. A caution must here be given not to look down the chimney more than necessary, and to do so quickly, since when a tube gives, serious injury to the sight would result to anyone caught. Passage between the ignition tube and the interior of the cylinder is opened and shut by the movement of an ignition valve, this valve being operated by a lever which is acted upon by a double cam on the screw shaft. When the piston is compressing the mixture, it is necessary that the communication between the cylinder and the ignition tube should be cut off, therefore the lever lifts the ignition valve, which enters an aperture, thus effectively closing the passage. To provide for any leakage of gas that may take place from the cylinder, a hole allows it to escape into the atmosphere. When the compression is finished, the cam allows the lever to drop, hence the valve drops on to its seat; the passage is now open, so that a quantity of the compressed gas from the cylinder rushes through into the hot tube, which ignites it at once; the burning gas being forced back re-enters the cylinder and ignites the compressed charge. Two chambers are provided to receive the burnt products of the gas which fires the charge, to avoid as much as possible any choking up of the passages by residuum, etc.

Life of Tubes.—The ordinary wrought-iron gas-tube may be reckoned to last 30 hours, but this is not guaranteed, for some only last a few hours; on the other hand, some may last 60 hours. In running, it is advisable to change them every 24 hours of actual work. If the gas-engine is not doing any particular work, when stopping would not signify much, such as charging accumulators in the daytime, etc., then the tube might be left in until it burnt out. Lately, Messrs. Crossley have put on the market a greatly improved tube; it is made of steel with some alloy, the exact composition of which is kept secret, and will last 12 months. Concerning these tubes, the author has just been informed by

Messrs. Crossley that one of their customers brought them a
tube which had been in use from January to October, and
was still in good condition after working 10 hours per day.
The superiority and advantages of these prepared steel tubes
over the wrought-iron tubes is thus evident, since it causes
great inconvenience and annoyance when the electric light
goes out suddenly through the failure of the ignition tube.
The cost of these steel tubes is high, 12s. 6d. each, and 15s.
with holder, whereas the iron ones are only 2½d. each ; in
addition, the steel tubes must be handled with great care, as
they are extremely fragile, and the slightest knock would
fracture them.   But, as said before, where it is of great importance for the gas-engine to run without danger of stopping,
the high cost of the steel tubes is a trifle in comparison.

Steady Running, etc. — The latest and most improved
form of Crossley's high speed gas-engine for electric lighting
is fitted with their patent controlled tube ignition.   These
engines give a most remarkable steady drive equal to a steam-
engine : this is owing to the two very heavy wheels on the
engine, either of which can be used as a driving wheel ;
these wheels are also loaded with a mass of metal near their
axis.   Another reason for their steadiness is their high
speed—250 revolutions per minute—so that two such heavy
wheels rotating at this speed possess an enormous amount of
momentum, and, therefore, the explosion of the charge does
not tend to jerk the speed.

Governors.—A gas-engine consumes gas exactly in proportion to the load, and this is effected by the aid of a governor.
In the case of the above type, on the screw shaft, which works
the valves, there is another cam, having three grooves cut in
in—each groove promoting a different action.   The governor
as it rises and falls in response to the load actuates a lever
which shifts to and fro, shifting a wheel into one of the
grooves, and thus the revolving cam moves the lever to a

greater or less extent. This movement of the lever works
the admission-valve, a strong spring attached to the lever
tending to hold the lever down, and so keep the valve closed
when the cam is not acting. By this arrangement the
amount of gas admitted into the cylinder is in proportion to
the power developed.

Advantages of the Gas-Engine.—Gas-engines are greatly
in demand where small motive power is required, and their
advantages on this score are not to be denied. They can be
put down anywhere where a supply of gas is to be had, can
be started in 10 minutes, stopped in one minute, and require
very little attention—very different from the constant watching which a boiler and steam-engine necessitate ; no flue or
chimney shaft has to be built, and the engine, having its
furnace in its cylinder, occasions no dirt or trouble in keeping
in order. It is not necessary to have a man on purpose to
run a gas-engine, except very large sizes, or where the plant
is one of great importance, say in a theatre, etc., where
great, if not serious, inconvenience would occur to the public
in the event of anything going wrong. In country houses
where they run an electric plant by a gas-engine, the plant
is usually placed under the care of the gardener or other
ground man, and who only needs to look in every hour to oil
up and see that everything is right ; in a number of places
the lights are run from accumulators or storage batteries,
the gas-engine running only during the daytime to charge
them from the dynamo, the lights being then disconnected.
Special oiling arrangements can be supplied with these
engines for this kind of work, so that they can be run for
10 or 12 hours without attention.

Running Cost of Gas-Engines.

An investigation will now be made respecting the probable
cost of running a gas-engine ; one of 35 b.h.p. will be

chosen, so that it will be about the same size as the 40-b.h.p. steam-engine, whose cost was given on page 70, the maintenance cost being divided up in the same way as with the steam-engine. A comparison of costs between gas and steam power can thus be made. The outlay for a gas-engine yielding about 35 b.h.p., including foundations, water-tank, pipes, and fixing in working order, may be put down at £280. The running cost will be divided into two parts, as was done when estimating the running cost of a steam plant.

Therefore we have—

TABULATION 14.

A.—INVARIABLE COSTS.

	B.H.P. per annum.	B.H.P. per hour.
Interest on outlay, £280 at 4 %	£0 6 4	·025 penny.
Depreciation ,, ,, 7½%	0 12 0	·048 ,,
Repairs ,, ,, 5%	0 4 8	·018 ,,
Attendance, two hours' time per day	0 8 7	·034 ,,
Total	£1 11 7	·125 penny.

B.—VARIABLE COSTS.

20 cubic feet of gas per hour for 3,000 hours, gas being 2s. 9d. per 1,000 cubic feet	£8 5 0	·66 penny
Oil, water, waste, and sundries	1 0 0	·08 ,,
Total	9 5 0	·74 penny

Giving total cost per brake horse-power per annum = £10. 16s. 7d.
And ,, ,, ,, ,, hour = ·865 penny

As with the steam-engine, let the gas-engine only work for 1,500 hours per annum, and be used for running electric lights from dusk to 10 p.m., another comparison can then be obtained respecting running costs, the working hours in this latter case being only half the former. The first part (A) of the cost will remain constant, while the second part

(B) will vary in proportion ; since 1,500 is half of 3,000, therefore the variable costs will be £9. 5s. ÷ 2 = £4 12 6
and the total will be £4. 12s. 6d. + £1. 11s. 7d. = 6 4 1
Giving the cost of 1 b.h.p. per annum .... = 6 4 1
„ „ „ „ hour .... = 1 penny.
Following up the comparison of cost of gas *versus* steam, we will now work out the probable cost of a 9-b.h.p. gas-engine, this small size being taken to compare with the 10-b.h.p. steam-engine, allowance being made for the lower efficiency, as was done with the small steam plant. The outlay for a 9-b.h.p. gas-engine, high speed, and designed for electric light plant, may be put down at £160, everything complete ; the running costs being as follows, the engine working 3,000 hours per annum :

TABULATION 15.

A.—INVARIABLE COSTS.

	B.H.P. per annum.	B.H.P. per hour.
Interest on outlay, £160 at 4 %	£0 14 3	·057 penny.
Depreciation „ „ 7½%	1 6 8	·100 „
Repairs „ „ 3 %	0 10 8	·043 „
Attendance, two hours per day	1 13 4	·133 „
Total	£4 4 11	·333 penny.

B.—VARIABLE COSTS.

	B.H.P. per annum.	B.H.P. per hour.
27 cubic feet of gas per hour for 3,000 hours, gas being 2s. 9d. per 1,000 cubic feet	£11 2 9	·890 penny.
Oil, water, waste, and sundries	1 5 0	·100 „
Total..................................	£12 7 9	·990 penny.

So that total cost per brake horse-power per annum = £16 12s. 8d.
And „ „ „ „ „ hour = 1·33 pence.

This is just over 50 per cent., or half as much again as the cost per brake horse-power of the large gas-engine ; or by spending two and one-half times as much per annum for a large plant, four times the power can be obtained, since 40 × £10 = £400, and 10 × £16 = £160.

Letting the small gas plant run for 1,500 hours per annum, the figures work out as follows: We have £4. 4s. 11d., which remains constant, and £12. 7s. 9d., which must be divided by 2, because the values of B are in proportion to the hours. So the total costs will be £4. 4s. 11d. + £12. 7s. 9d ÷ 2, or £4. 4s. 11d. + £6. 3s. 11d. = £10. 8s. 10d. per brake horse-power per annum, or 1·66 penny per brake horse-power per hour.

When motive power is required for running electric lights on business premises, the time from dusk to 10 p.m. (which is 1,456 hours per annum including Sundays, and 1,248 hours per annum excluding Sundays) is in most cases too long, and calculations for much shorter hours might be acceptable. From dusk to 8 p.m., including Sundays, means about 742 hours per annum, or 636 excluding Sundays, and from dusk to 8.30 p.m., excluding Sundays, signifies close on 786 hours per annum. Taking half of 1,500, we have 750, and this number of hours will be near enough to judge of the cost of running from dusk to 8.30 p.m., excluding Sundays.

The results are as follows, 750 hours running :

Large steam plant
per b.h.p. .. ... = £6 5 0 per annum, and = 2·00d. per hour.
Small steam plant
per b.h.p. ...... = 14 17 3 per annum, and = 4·75d. per hour.
Large gas plant
per b.h.p. ..... = 3 17 10 per annum, and = 1·24d. per hour.
Small gas plant
per b.h.p. ...... = 7 6 10 per annum, and = 2·35d. per hour.

In Tabulation 16, given below, the results of running costs, as calculated out for steam and gas engines, are collected together, so that the difference existing between the two can be seen at a glance. The large steam-engine gives 40 b.h.p., and the large gas-engine gives 35 b.h.p., these two sizes being sufficiently near for comparison ; same with regard to the small engines, where the steam-engine is 10 b.h.p., and the gas-engine 9 b.h.p. The columns denoted by A give the price of 1 b.h.p. per annum, whilst the columns denoted by B give the price of 1 b.h.p. per hour in pence.

TABULATION 16.

Large Plant.

Hours per annum.	Steam.		Gas.	
	A.	B.	A.	B.
3,000	£10 0 0	·800 or $\frac{4}{5}$	£10 16 7	·865 or $\frac{7}{8}$
1,500	7 10 0	1·200 or 1$\frac{1}{5}$	6 4 1	1·0 or 1
750	6 5 0	2·0 or 2	3 17 10	1·24 or 1$\frac{1}{4}$

Small Plant.

Hours per annum.	Steam.		Gas.	
	A.	B.	A.	B.
3,000	£19 11 0	1·56 or 1$\frac{1}{2}$	£16 12 8	1·33 or 1$\frac{1}{3}$
1,500	16 8 6	2·63 or 2$\frac{2}{3}$	10 8 10	1·66 or 1$\frac{2}{3}$
750	14 7 3	4·75 or 4$\frac{3}{4}$	7 6 10	2·35 or 2$\frac{1}{3}$

Some interesting results are shown by this tabulation. It will be noticed that when running for 3,000 hours per annum the steam-engine and gas-engine are about on a level for cost, and that as the running hours are shortened so the cost of a gas-engine becomes less than that of a steam-engine. Coming to the smaller plant, we see that gas has a most decided advantage, even for long hours, and for short hours it is only half the cost of steam. We are thus led to the following conclusion.

With engines of about 40 b.h.p. running 3,000 hours, the
cost for steam and gas are about equal ;
As the power decreases, so gas will be less than steam ;
As the power increases, so steam will be less than gas ;
As the hours decrease, so gas will be less than steam ;
As the hours increase, so steam will be less than gas.

Dowson and other Gas Producers.

A gas producer is an apparatus for manufacturing explosive
gas or vapour. The fuel used may be divided into two
classes : (1) coal, wood, etc. ; (2) petroleum and other oils,
fats, etc.

The best-known gas producer used for generating gas for
driving gas-engines is Dowson's, the gas produced being
named Dowson gas. This gas is made in the following way :
Coal, or other suitable fuel, is burnt in a steam producer ;
the steam is mixed with air, and the mixture forced through
an incandescent mass of burning fuel placed in the generator.
The steam is decomposed into its constituent parts—namely,
hydrogen and oxygen. The hydrogen passes off, while the
oxygen from the steam and from the air combines with the
carbon to form carbonic acid gas (CO_2), this afterwards,
mixing with more carbon, becomes reduced to carbon
monoxide (CO), so that the gas produced consists of carbon
monoxide, together with hydrogen, and small quantities of
CO_2, nitrogen from the air being present in large quantities.
Owing to the large quantity of carbon monoxide gas that is
contained in Dowson gas, this gas must be used with great
caution, since CO is a deadly poison, and an atmosphere
containing, say, 2 per cent. of it would endanger life. The
plant, however, is made very secure, and although used a
great deal very few fatalities have occurred.

The heat of combustion of Dowson gas is one-quarter of
that given by an equal volume of coal gas, consequently four

·times as much is required to drive gas-engines : so that where 20 cubic feet of coal gas is consumed per brake horse-power more than 80 cubic feet of Dowson gas is required. The plant takes up very little room, and gas is manufactured at a very rapid rate. For example, a ground space of about 10ft. square would accommodate a plant sufficiently powerful to supply gas to a gas-engine giving 45 b.h.p. Comparing this with the space a steam-boiler takes, it is about on a par.

From tests which have been made, it is found that using coal to produce Dowson gas to drive a gas-engine, only 1·3lb. of coal are consumed per brake horse-power per hour. This test was for an engine giving 7 b.h.p. A more recent test, where a Crossley Otto gas-engine developing 170 i.h.p. was driven by Dowson gas, gave a consumption of coal of ·883lb. per indicated horse-power hour, and only 52lb. of water for everything—a most astonishing performance ; this works out to ·08, or about $\frac{1}{12}$ of a penny per brake horse-power hour, using coal at 15s. per ton.

As an example of gas made from oils, fats, etc., we will take Mansfield's oil-gas apparatus. In this apparatus, the oil, or fatty matter, is vaporised in a retort heated by a fire of coal, wood, or other suitable fuel ; the gas given off is passed through water to abstract all tar products, and is then purified by being passed over a mixture of slaked lime and sawdust. The kind of oils mostly used are those known as "intermediate": they have no unpleasant smell. It is found that 1,000 cubic feet of gas can be made from 10 gallons of oil, or one gallon will supply 100 cubic feet of gas. For large-sized gas-engines the consumption of oil gas may be taken as about 10 cubic feet per brake horse-power per hour. Oil costs, say, 4½d. per gallon, and the cost of fuel in the furnace is about 1d. per 100 cubic feet of gas produced, and since one gallon will produce 100 cubic feet, hence total cost will be 5½d., or, say, 6d. per 100 cubic feet, allowing

G

10 cubic feet for 1 b.h.p., this comes to ·6, or $\frac{2}{3}$ of a penny per brake horse-power per hour.

Petroleum Oil-Engines,

Of late years petroleum oil has provided a most important and extensive source of fuel, and entirely supplies the place of coal in districts where that fuel is scarce and oil is plentiful, and every year this oil becomes more used. In South Russia, where the greatest petroleum wells in the world exist, the refuse of this oil is used in the place of coal in the furnace of the locomotives, the oil being stored in an oil-tank for the purpose—it is also used for steamers; 1lb. of petroleum oil will give out 50 per cent. more heat energy than 1lb. of coal, so that to give the same heat energy, oil fuel is 33 per cent. less in weight than coal fuel. In addition, it only occupies about one-half the bulk that coal does.

Another way of using petroleum for motive power is by vaporising it, and then exploding the oil-gas, as in Mansfield oil-gas producers. But there is also a third way, and this is, perhaps, the most important, or at all events will become so, in every probability ; this is by using the oil both as a fuel and motive force in the cylinder of the engine. This is illustrated by Priestman's petroleum oil-engine.

The engine is very compact and entirely self-contained. The foundation of the engine contains a water tank, supplying water to the cylinder jacket ; this does away with a separate tank and piping. Underneath the engine, by the crank, is situated the oil-tank, the oil for a day's run being usually put in, while under the cylinder is placed the vaporiser ; a side shaft works two pumps, the larger one being for compressing air, and the smaller one for circulating water round the cylinder. The working of the engine is as follows : the air-pump forces air at a pressure of about 7lb. per square inch on top of the oil; a stream of oil is then forced along a pipe, and another stream of air along another pipe; both join and lead into the vaporiser, and then the air breaks the oil into a spray; whilst passing through the vaporiser the mixture of oil and air is heated by the exhaust gases from the cylinder. As the piston makes its forward stroke, it sucks in after it, by means of a suction valve, a quantity of oil spray and air. After compression this mixture is fired by an electric spark. The wires from an induction coil are led into the cylinder, the spark jumping across the gap between the two wires; the electric circuit is closed by the side shaft moving backwards and forwards, and so bridging across two terminals, by means of a contact-piece. The burnt gases driven out from the exhaust-port are utilised in heating the mixture passing through the vaporiser. The oil, when exploding, has the valuable property of lubricating the piston. Common petroleum oil is used, which can be obtained almost anywhere; this oil is particularly cheap in seaport towns, therefore oil-engines should thrive in those places; an average price is 5½d. per gallon.

The amount of oil required to produce 1 b.h.p. per hour is proved to be less than 1¼ pints of oil; this will cost, therefore, ·86d., or $\frac{7}{8}$d. One gallon of common petroleum oil, specific gravity ·800, weighs about 8lb. And since 1lb. of petroleum is equivalent to 1½lb. of coal in heat energy, therefore 1¼ pints of oil = 1¼ × 1½ = 1·9, or nearly 2lb. of coal; this is equal to 14,500 × 2 = 29,000 heat units to produce 1 b.h.p. per hour. So that the efficiency of the oil-engine is about equal to the very best steam-engine, as it utilises about 10 per cent. of the heat energy of the oil, and has half the efficiency of the gas-engine using 22·5 cubic feet of gas per hour.

The way to start the Priestman oil-engine is as follows.

First, the vaporiser must be heated by means of a hand oil lamp (a few minutes suffice for this), like heating the ignition tube of a gas-engine. The air-pump is then worked by hand by a lever for that purpose, placed at the crank end of the engine (half-a-dozen strokes is sufficient for this), two or three turns of the flywheel, a couple of explosions, and the engine is off.

Common petroleum or lamp oil is perfectly safe to store and handle, since if a lighted match be dropped into a vessel full of this oil the oil will neither explode or burn, and the light will be quenched as if it had fallen into water.

The following tabulation gives the efficiency and cost of fuel per brake horse-power per hour for different motors :

TABULATION 17.

Motor.	Fuel.	Per B.H.P. per hour.	Heat units.	Efficiency.	Cost B.H.P. per hour.
Steam.	Coal.	4lb.	58,000	4·5%	·32 penny.
Gas.	Gas.	22·5 cub. ft.	15,000	17·2%	·75 ,,
Oil.	Oil.	1·25 pints.	27,550	9·3%	·86 ,,

Coal being at 15s. per ton, gas at 2s. 9d. per 1,000 cubic feet, and oil at 5½d. per gallon.

Water Power.

This is a source of power that exists to a very limited extent in England, because the rivers are not of a rapid nature, and there are very few waterfalls of any size. In a country like Switzerland water power is abundant, due to the swift mountain torrents and waterfalls that are met with everywhere. It is, therefore, only natural that the Swiss engineers are famous for the turbines manufactured by them, and their experience in utilising this source of power. Water power, for motive purposes, is used all over the world. whenever it can be obtained, and proves a most valuable and cheap source of power where coal or any such fuel is scarce and expensive. The development and practical application of transmission of power by electricity has made a good source of water power of still greater importance and value, since by electricity the energy of the water, instead of running to waste, can be transmitted a number of miles and there utilised, subject to a loss depending on the distance and the method employed.

A moving body of water may be said to deliver its power in two general ways, in a practical sense : first, by momentum ; second, by weight. A rapid stream explains the first, a waterfall explains the second. All water-motors may be divided into two classes : (I.) Waterwheels (horizontal axis) ; (II.) Turbines (vertical axis).

(I.) *Waterwheels.*—This class of water-motor may be again divided into three types—namely, (*a*) undershot, (*b*) overshot, (*c*) breast.

The undershot waterwheel is the oldest form of any device used to obtain power from a moving body of water. Figs. 3 and 4 show an undershot wheel. The water flows in what is named a " race," the part in front of the wheel being called the "head race," whilst the part behind the wheel is called the " tail race." In the head race the water is flowing at its normal rate, and so is in a condition to do work. Upon reaching the wheel, which offers an obstruction, the water forces the floats in front of it, and so turns the wheels ; the water now has parted with more or less of its power, and so continues its way at a slow rate into the tail race.

The floats are pieces of board about ¾in. to 1½in. thick ; these have the same width as the rim of the wheel, and are supported by stout wedges of wood placed behind them. The diameter of the wheel determines the number of arms. With a diameter of, say, 12ft., there would be about six arms, the floats being placed about 15in. apart. That portion of the

race just underneath the wheel is curved out, so as to fit
the wheel, the race being made the same breadth as the wheel.
The average depth of water should be about 7in. or 8in., and
this depth is regulated by a gate, fixed in a slanting position ;

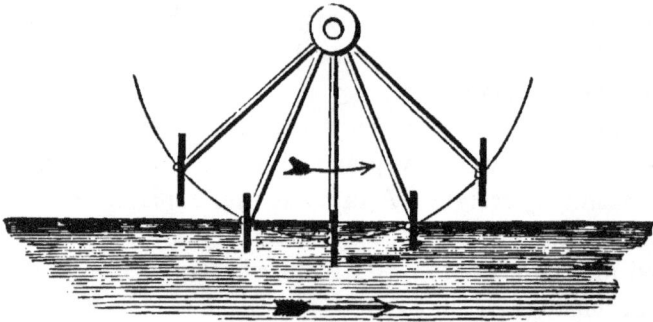

FIG. 3.

by raising or lowering this gate, a greater or less quantity of
water can be admitted through the open space between the
bottom of the gate and the bed of the race. The angle of
slant for this gate must be such that there is a very small

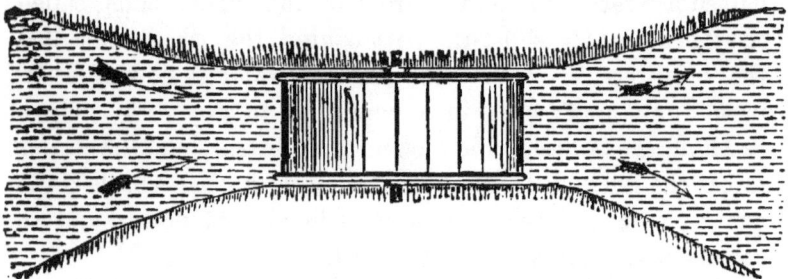

FIG. 4.

distance between the opening and the tips of the floats. By
this means the water impinges right on to the floats immediately on getting through the gate, and so power is not lost.

Undershot wheels should be used where the streams are

rapid, contain large bodies of water, and are not subjected to much rise or fall, for in the event of a flood the wheel would become " drowned," and its power crippled.

The great advantages which an undershot wheel possesses are cheapness in first cost and simplicity of construction, so that it would not be very difficult for anyone to make one for themselves, and they would find that the cost would not be much.

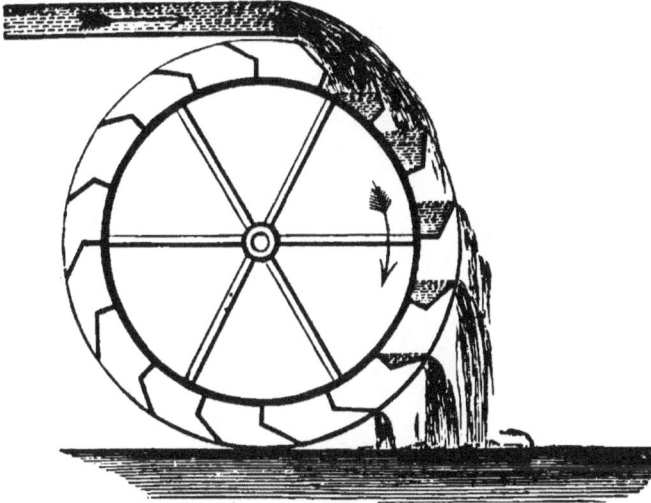

FIG. 5.

The. Overshot Waterwheel receives its power as much from weight as from momentum of the water. In place of the " flume " in which the undershot wheel receives its power, the water is delivered to the overshot wheel from above, so that it falls just beyond the top centre into buckets fixed to the periphery of the wheel. Fig. 5 gives an illustration of an overshot wheel, and the way in which water is delivered to it. A wooden trough made of planks delivers the water ; at the end of the trough a thin plate of iron is fixed on the bottom of the trough and projects out,

forming a lip; this lip projects a little beyond the
vertical diameter of the wheel, the water as it rushes
out leaps off the lip, and so is carried a short distance
forward, and, therefore, falls on the buckets a little beyond
the top of the wheel. A sluice is fixed at the entrance to
the trough, and so the water can only enter in a thin stream.
This type of wheel can only be used where there is a good
fall of water—quantity of water does not much signify,
provided there is a fall. It may be reckoned that an overshot wheel requires a fall not less than about $1\frac{1}{4}$ times the
diameter of the wheel.

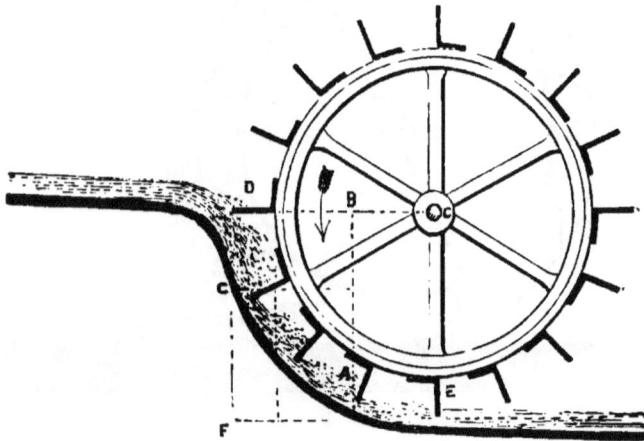

FIG. 6.

The Breast Waterwheel is a combination of the undershot
and overshot, and is probably more used than either of the
others. As with the undershot, a flume of masonry is built
for the race ; the walls of the flume are so made that the
wheel will just fit in, there being scarcely any space left
between the floats and the walls. This is in order to
prevent any water from escaping past the floats. Fig. 6
shows how the water is delivered to a breast wheel. A

curved piece of iron fixed at the entrance to the flume is called the " guide bucket."

For high speed, use wide floats and small diameter.

For low speed, use narrow floats and large diameter.

The following is the efficiency of the three types of waterwheels :

Overshot utilises.................. 50 to 70 per cent.

Breast utilises..................... 45 to 50 ,,

Undershot utilises .... 27 to 30 ,,

The Pelton Waterwheel.—This motor is a great improvement on the ordinary wooden waterwheel, being built of iron and encased by an iron shell, the water being led into it by an iron delivery-pipe ; a regulating wheel valve regulates the amount of water admitted, and so the power of the motor can be varied. These motors have lately been provided with a throttle-valve and centrifugal governor, so that the speed can be regulated when the load varies.

(II.) *Turbines.*—This class of water-motor may be divided into two types—namely : (*a*) outward flow ; (*b*) inward flow.

When the power is used for driving dynamos, and running electric lights direct from the dynamo, then a turbine should be used, because a waterwheel cannot be depended on to run steadily ; an ordinary waterwheel with its irregular running is quite good enough for driving a dynamo when it is only used for charging accumulators, because it is not absolutely necessary to have a constant speed, and a variation of speed will not matter provided it is within reasonable limits. Of course a great deal depends on the nature of the water power. In districts where it remains fairly constant in power the lights can be run successfully by a waterwheel, provided that some regulator is used, such as putting an electrical regulating device across the main terminals, of the dynamo so as to throw resistance in and out of the shunt coils.

The principle underlying the action on which turbines work, may be termed to be the effect of a reactionary force. For an example of what is meant by this definition, it will be remembered that the first instance in which steam was employed to promote motion, and consequently yield power, is generally ascribed to Hero of Alexandria, who partly filled a metal sphere with water and attached two curved spouts diametrically opposite to each other ; upon boiling the contained water, steam issued from the spouts, one jet rushing upwards and the other jet rushing downwards, and the sphere, being pivoted, revolved. This illustrates to some degree the meaning of reactionary forces. The following explanation given by Bodmer gives this principle of reaction as regards water : " The construction of turbines is based upon the fact that when a mass moving in a given direction with a given velocity is impelled to change this direction, force is required to effect this change. The intensity of the force necessary is obviously dependent on the extent to which the mass is deflected from its original course. In a turbine a jet of water is deflected by being brought into contact with a curved vane, which, preventing further progress in the initial direction, compels the water to follow its surface. Owing to the resistance offered by the water to this compulsion, a reactionary force is exerted on the vane which is employed driving the turbine wheel." Fig. 7 illustrates how the above reactionary force is practically applied, where, instead of the water being turned abruptly from a vertical to a horizontal course, the arms make a curve as seen in the section.

In outward-flow turbines the water enters at the centre, and leaves at the circumference, the guides being placed inside the wheel. In inward-flow turbines the converse holds—that is, the water enters at the circumference and leaves at the centre, the guide blades being fixed round the periphery of the wheel. A good turbine should give out from 70 to 75 per cent. of the available power, so that they are far more efficient than waterwheels.

Calculation of Water Power.—After the explanations given as to the meaning of horse-power and how it is calculated in a steam-engine, it will not be very difficult to do the same for water power. Quantity of water is usually stated in gallons, but in calculating power the water must be measured in pounds weight, because the term gallon is no use, and has to

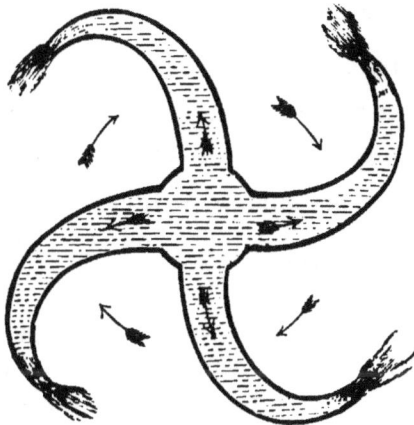

FIG. 7.

be changed into weight. When the water yields power, by virtue of its dead-weight, as when it falls freely from a height, then the theoretical power it gives out just upon reaching the ground is measured simply by the weight in pounds per minute that flows, and the distance in feet that it falls. Now one cubic foot of water weighs 62·5lb., therefore one cubic foot per minute falling 1ft. will yield 62·5 foot-pounds per minute, and since 33,000 foot-pounds make 1 h.p., therefore 528 cubic feet of water per minute, or 8·8

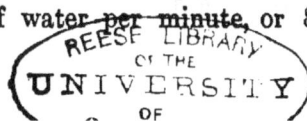

cubic feet per second, falling freely 1ft., will produce 1 h.p. In order to obtain the number of cubic feet of water, the velocity of the water must be measured as near to the fall as possible. The average depth and width of the stream multiplied together give the average sectional area of the water that is flowing, therefore multiplying this by the velocity in feet per minute will give cubical amount of water that flows per minute.

Example.—Suppose a stream, having an average depth of 5in. and an average width of 9ft., has a free fall of 13ft. ; find its horse-power, the velocity just above the fall being, say, five miles per hour.

The sectional area $= 9 \times (5 \div 12) = 3.75$ square feet ;

Five miles per hour $= 440$ft. per minute ;

∴ quantity of water per minute $= 3.75 \times 440 = 1,650$ cub. ft.

∴ weight of water per minute $= 1,650 \times 62.5 = 103,125$lb.

Hence, power developed $= \dfrac{103,125 \times 13}{33,000} = 40.6$ h.p.

In the above case we have measured the power given out by the water by a certain weight of water in pounds falling a certain distance in feet in a certain time—one minute—that is to say, in foot-pounds per minute.

Now turn attention towards measuring the power of water developed, not by weight, but by momentum—momentum signifying the amount of motion possessed by a moving mass.

When a mass of water is travelling along at a certain velocity, the amount of power it would yield upon being suddenly stopped, would be equal to the amount of power that was given to it in order to give it that velocity. The simplest way to find out the amount of power possessed by a moving mass, is to calculate out what vertical distance the mass would have to fall, acted upon by gravity, in order to acquire that velocity, and then to proceed as usual. A body falling freely is constantly accelerating its velocity or rate of motion. Thus it starts from rest, having a velocity of 0, and goes on increasing in velocity until at the end of one second of time its velocity is 32·2ft. per second, and its *average* velocity throughout the space of that second is naturally (0 + 32·2) ÷ 2 = 16·1ft. per second and the distance fallen is 16·1ft. In the next interval of time, its *initial* velocity is already 32·2ft. per second, while its final velocity is 64·4ft. per second at the end of the second second ; hence its *average* velocity during the two seconds is (0 + 64·4) ÷ 2 = 32·2ft., and so on until at the end of 12 seconds its *final* velocity would be 386·4ft. per second, its average velocity (0 + 386·4) ÷ 2 = 193·2ft. per second, and therefore the total space covered by the falling body would be 193·2 × 12 = 2,318·4ft., so the law is—

Space = {0 + (32·2 × time)} ÷ 2 × time, or $s = 16·1\ t^2$,

where s = space in feet, and t = time in seconds.

In words, this means that the distance in feet a body will fall is obtained by multiplying the square of the time in seconds by 16·1.

Example.—A stone falls down a precipice ; at the end of four seconds it is seen to strike the surface of a pool of water at the bottom ; how deep is the precipice ? Applying our rule, we have

$$s = 16·1\ t^2 = 16·1 × 16 = 257·6\text{ft.}$$

We have now obtained the law connecting space and time for falling bodies, and from this we have to deduce another law—one connecting final velocity and space—so that when a mass is moving at a certain velocity we can find out what distance it would have to fall in order to acquire that velocity. This law would then enable us to measure the energy of the moving mass, because the weight of the mass multiplied by its distance of falling will give us the foot-pounds of energy it would develop upon being suddenly stopped.

The following is the deduction : Indicating final velocity by v, $v = 32 \cdot 2 \ t$; therefore by dividing the velocity of the mass by $32 \cdot 2$, we obtain the number of seconds it would have to fall to acquire that velocity; and knowing this, the space covered can be easily calculated from the previous law, $s = 16 \cdot 1 \ t^2$, since by substituting the value of t, we get $s = 16 \cdot 1 \times (v \div 32 \cdot 2)^2 = \text{space}$. This is worked out in the following way :

$$v = 32 \cdot 2 \ t \ ; \ \therefore \ t = v \div 32 \cdot 2 \ ; \ \therefore \ t^2 = (v \div 32 \cdot 2)^2 \ ;$$

$$s = 16 \cdot 1 \ t^2 \ ; \ \therefore \ s = 16 \cdot 1 \times (v \div 32 \cdot 2)^2 = \frac{v^2}{64 \cdot 4}$$

Our law, then, is this : Square the velocity of the mass expressed in feet per second, then divide by $64 \cdot 4$. The answer gives the distance in feet the mass would have to fall, acted on by gravity, to acquire that velocity. Hence the stored energy possessed by a moving mass $= \dfrac{(\text{velocity})^2}{64 \cdot 4} \times$ weight in pounds.

This law is usually expressed by the formula, $\frac{1}{2} \ m \ v^2$, where m signifies the mass of the substance   The " mass " is obtained by dividing the weight in pounds by the value of " g," which in England is $32 \cdot 2$.

Example.—Suppose a swift stream of water is rushing down an incline at the rate of seven miles per hour, its average depth being 9ft., and its width 25ft. ; what horse-power will the stream produce ?

Seven miles per hour = 10·26ft. per second ;

Cross-section of channel = 25 × 9 = 225 square feet ;

Mass of water flowing by per second = 225 × 10·26 = 2,308·5 cubic feet ;

Weight of water flowing by per second = 2,308·5 × 62·5 = 144,281·25lb. ;

$$\therefore \frac{(\text{velocity})^2}{64\cdot4} \times \text{weight} = \frac{(10\cdot26)^2}{64\cdot4} \times 144,281\cdot25$$

$$= 231,615\cdot6 \text{ foot-pounds per second.}$$

Hence horse-power $= \dfrac{231,615\cdot6}{550} = 421\cdot1$

The following tabulation gives a number of values of velocity worked out in their equivalent values of distance :

V = velocity = number of feet per second.

S = space or head in feet.

TABULATION 18.

V	S	V	S
1	·01	20	6·2
2	·06	30	13·9
4	·24	40	24·8
6	·57	50	38·8
8	1·00	60	55·0
10	1·5	80	100·0

Feet per second = 1½ times the number of miles per hour (2¼ per cent. too high).

Miles per hour = two-thirds the number of feet per second (2¼ per cent. too low).

The preceding values are plotted out in a curve, as shown in Fig. 8, where velocities are plotted along the vertical axis, and the spaces along the horizontal axis.

Since one cubic foot of water weighs 62·5lb., therefore

the energy of one cubic foot of water moving with V
velocity is

$$\frac{V^2}{64\cdot4} \times \frac{62\cdot5}{1} \text{ foot-pounds.}$$

Multiplying this by the number of cubic feet of water that
flow by per second, will give the foot-pounds developed per

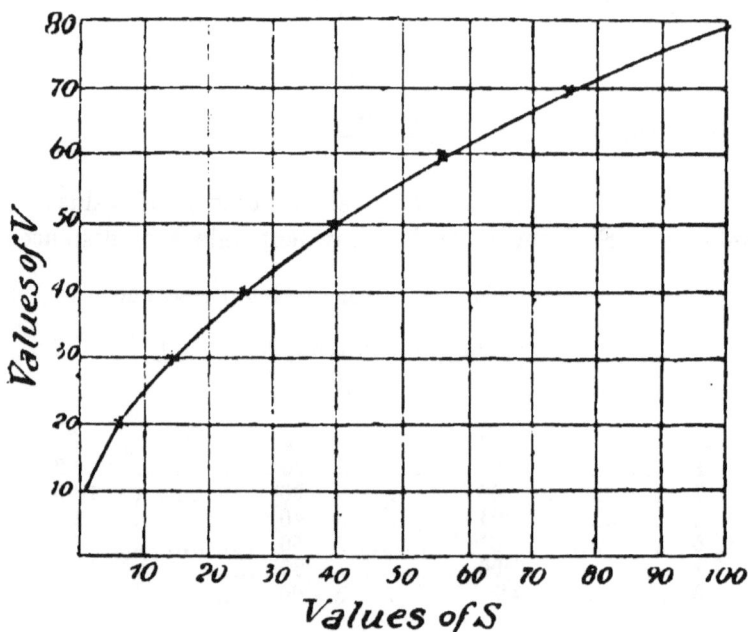

FIG. 8.

second by a volume of flowing water ; 550 foot-pounds per
second are equivalent to 1 h.p., and since 62·5 is sufficiently
near to 64·4 for approximate results, these two factors can
be omitted.   Now, volume in cubic feet per second is the
product of cross-sectional area in square feet and velocity
in feet per second ; hence this formula may be written

$$\frac{V^2 \times \text{area} \times V}{550} = \frac{V^3 A}{550} = \text{H.P.},$$

where V = velocity in feet per second,

and   A = cross-sectional area of stream in square feet.

We can thus obtain the horse-power of a flowing stream of water by the use of the following rule for approximate results only : Cube the velocity of the stream, expressed in feet per second, multiply this by the average cross-sectional area expressed in square feet, and then divide by 550.

The above rules for calculating water power only give theoretical results, since there are several things which must be taken into consideration in obtaining the actual power— for example, the depth must be very carefully noted in several places, and only the mean or average taken. Then again it must be remembered that the water at the bottom and at the bank sides flows slower than that in the central parts of the stream. This is due to the friction existing between the earth and the water near it, as everyone knows that in rowing against the stream a boat should hug the bank.

CHAPTER III.

PRACTICAL LAWS OF ELECTRICITY AND MAGNETISM.

Sources of Electricity—Water Analogy—Electrical Units. Ohm's Law—Calculation of Resistance—Conductors and Insulators—Heating Effects of the Current—Simple and Divided Circuits—Magnets and Magnetism—Magnetic Properties of Iron—The Magnetic Circuit.

Sources of Electricity.

It is not a very easy matter to explain correctly the elementary laws of electricity when this explanation is confined to simple and analogous language, debarred from scientific or mathematical proof, but it is necessary that those interested in electrical affairs should possess some slight knowledge of these laws and the meaning and import of electrical phraseology used in connection with the practical applications of electricity. Everybody knows that a steam-engine works by steam pushing a piston up and down inside a cylinder, but very few people indeed have anything beyond the faintest inkling as to why an electromotor will rotate when a current of electricity is passed through it, and it is not too much to say that nearly all the rest do not know what an electromotor is, in the first place. The electric light is spreading so rapidly and the industry becoming of such universal importance that such common terms as "watt," "volt," "ampere," and the like should be known to every educated person. Consumers of electricity are left supremely ignorant when they are informed that the price of 1,000 watt-hours is 7d. or 8d., as the case may be, and that such lamps are run at a pressure of 100 volts, and yet very little electrical knowledge is required to understand these elementary terms. In the next generation, when electricity will occupy a large amount of public attention, these simple expressions will then of necessity become generally known and understood, and, therefore, will probably take their place in the dictionary along with technical terms of other subjects which usage has embodied in our language, and then " units " of electricity will be spoken of in the same way as horse-power of steam-engines or cubic feet of gas.

The motion of any body is due to the action of some force, and the movement of a quantity of electricity, such as a current of electricity flowing along a conducting path or circuit, is due to the action of what is technically called the " Electromotive Force," or " E.M.F.," known briefly in practical electric work as " electric pressure," and often simply as "pressure." This E.M.F. can be promoted in several ways.

1. By rubbing a vitreous substance (like glass) with silk, or a resinous substance (like ebony or sealing-wax) with flannel. The phenomena arising from rubbing amber were known a long time ago, because it was known by Thales of Miletus (B.C. 640-548), and the word "electricity," expressive of this phenomena, seems to have been derived from " elektron," the Greek word for amber. It was Dr. Gilbert, the physician to our Queen Elizabeth, who discovered that resinous and vitreous substances shared this peculiar property with amber, and in his Latin treatise on the subject the word " electricita " is used to express this property. This kind of electricity has been named " Frictional or Static," the term " static " being employed because it is of a stationary nature, and exists as a charge on the surfaces of bodies. This charge of electricity may be of enormous E.M.F., but when the body holding the charge is discharged, the length of time during which a

H 2

current of electricity flows is so small that it may be called instantaneous.

2. By the chemical action of two kinds of metals, as in a primary or voltaic cell, so named after Volta, who first discovered this source of electricity in 1800.

3. By heating one of the two junctions that two different metals make when forming a circuit or closed path. This is called "Thermo-Electricity," and was first discovered by Seebeck in 1822. The E.M.F. that is produced in this way is only very feeble.

4. By moving a closed coil of wire in a magnetic field, as in the dynamo. This is named "Dynamic Electricity," because mechanical motion is used to produce the electricity ; hence the word "dynamo," signifying dynamic action.

5. By allowing stored chemical energy to do work and become converted into electrical energy, as in a "Storage Cell," or "Accumulator," called also a "Secondary Cell."

The last four sources are the only ones that will yield a flow of current for practical use. There are, however, a number of actions and effects on different bodies which will produce electrical phenomena, amongst which may be placed the following : Vibration, as when a metal bar, covered with an insulating material, is caused to vibrate ; rupture, or fracture, of certain materials ; percussion, when one body strikes another ; combustion ; crystallisation of certain chemicals ; pressure, etc. In addition may be mentioned animal electricity, as the torpedo, gymnotus, and other electric fish. Lastly, there is atmospheric electricity.

Of the four practical generators of electricity—namely, the Primary Cell, the Thermopile, the Dynamo, and the Secondary Cell—the last must be disregarded, because in the present way in which secondary cells are made, electrical energy has first to be put into them, and so converted into chemical energy, before they can be utilised to give out electrical

energy. Concerning the primary cell, this is too expensive
and impracticable, as shown in the paragraph on "Sources of
Power." As for thermopiles, attempts have been made to
construct these on a large scale, so as to make them powerful enough to run electric lights, but so far they are not
classed as being of practical use. It is possible that in the
future better things may be expected of this source of electricity, and it is interesting to note that, by means of a
thermopile, an example is afforded of generating electricity
direct from heat, and hence from burning coal, without any
intervening converters. To what extent this will become
practicable, it is impossible to say ; the subject is, however,
worth research.

Having disposed of three, we thus have only one practical
source of electricity available from which to obtain electric
light and power on a large and commercial scale, and that is
the "Dynamo."

Water Analogy.

The simple laws regulating the flow of electricity along a
conductor can best be made clear and comprehensible by
comparing them with those regulating the flow of water in
pipes. It must, however, be remembered that the analogy
between electricity and water cannot be strained and carried
too far. This water analogy is a time-honoured one, and
although modern theories and speculations concerning the
nature of electricity and its attendant laws as taught to-
day are different to what were taught in the old schools,
yet, for all we know respecting what electricity is, the
water analogy still holds good.

In the case of water a flow is produced by a difference
of level, consequently a head is created ; the greater the
head or pressure the quicker the flow, and the greater will
be the quantity of water that will flow through the conveying pipe. By decreasing the diameter of the pipe more

resistance will be offered to the flow, and by increasing the diameter less will be offered ; the pressure of the water, and hence its velocity, will gradually fall off along the length of pipe, and if the pipe be made very long the pressure will be reduced so much that the water at the far end will scarcely move along. We will now see how this can be applied to electricity flowing along a wire.

1. The pressure or difference of level of the water may be likened to the E.M.F., or " Difference of Potentials," that tends to move electricity ; the greater this is, the greater is the flow of electricity.

2. The size or diameter of the pipe may be likened to the diameter of the conducting path or circuit. Water flowing in a pipe produces friction, and hence heat, and this is energy wasted, so a current of electricity flowing along a wire, or other conducting path, produces what may be termed electrical friction, and this also produces heat, and hence wasted energy ; by making the conducting path of greater capacity, say, by increasing the diameter of the wire that forms the circuit, a greater current of electricity will flow.

3. The lengthening of the pipe may be likened to the lengthening of the conducting path or circuit ; if the wire composing the circuit be made very long, the electrical potential will fall along the circuit. In the case of water the effect of a long pipe means that a great quantity flows by at the beginning of the pipe, while scarcely any flows by at the end ; in the case of electricity this does not hold, for the effect of a long circuit on the flow of current is that very little current will flow, and that, however much it is, the amount of current that flows is the *same quantity per second at every point of the circuit.*

4. In the case of water, the effect of having a pipe of varying size or diameter at various points, is that at those

points where the diameter is small only a small quantity
of water can get through at a time, while at those points
where the diameter is large, a large quantity can pass :
but when the pipe narrows the velocity increases, and
when it widens the velocity decreases, consequently the
product of the sectional area and the velocity at any
point will be constant. Now, in the case of electricity, as
already stated, the current is always the same at every
point in the circuit, and the effect of having the con
ducting path narrow at some points and wide at others
would not cause the current to vary at these points.
What would take place is this : at the narrow points
more heat would be generated by the flow of current,
and at the wide points less heat would be generated.

5. Water, to be carried a long distance, must have very
great pressure, and the quantity must be necessarily small,
hence a long thin pipe of small diameter must be used.
Electricity, to be carried a long distance, must have
very great E.M.F., and the quantity must be necessarily
small, hence a long thin wire of small diameter must be
used.

In a water system, the initial or maximum pressure
decreases as the water flows on ; and in an electrical
system or circuit, the initial or maximum electrical
potential also decreases as the current flows along the
conducting path. In each case this is owing to the
resistance that is met with ; and the " difference of
pressure " and " difference of potential " that exists
between any two points of the two systems measures the
loss of pressure or potential—or the amount that is absorbed
in forcing the water or electric current through the
distance between these two points, and so overcoming the
mechanical or electric friction for that distance.

Water is a material body, whilst electricity is looked upon

as a peculiar state or condition of matter—what, we do not know. We even do not know in which direction a current of electricity really flows, although for convenience it is arbitrarily assumed that it flows from the point of higher potential to the point of lower potential. Furthermore, we do not know that its action is propagated inside the wire or not ; to every appearance its action takes place around the wire or conducting path, and the wire may be looked upon, not for the purpose of carrying the current, but of guiding it.

In a simple electric circuit there can only be one electromotive force, and this is called the initial or maximum difference of potentials ; but there can be an infinite number of difference of potentials. For example, a battery may have an E.M.F. of 50 volts—that means the maximum difference of potential is 50 volts—and allowing the current to flow from the positive plate to the negative, the potential will gradually fall all the way around the circuit, beginning at the positive plate and ending at the negative ; so that any two points in this circuit will represent a difference of potentials in proportion to the distance between, provided that the resistance of the external circuit is uniform throughout.

Electrical Units.

The practical unit of E.M.F. is named the volt, and is that amount of electrical pressure that will force a current of one ampere through a path having the resistance of one ohm. Volts are usually denoted by the letter E. Since the E.M.F. may also be looked upon as the maximum difference of potential in a circuit, hence the volt is the unit of difference of potential. Measured in absolute, or C.G.S. units, the unit signifies that E.M.F., or difference of potential, which is necessary to force a unit quantity of electricity per second from the point of higher potential to the point of lower potential, so that the current does one erg of work per second. This can be made clearer by referring to our water analogy. When a certain mass of water falls from a higher level to a lower level, then work is done by the mass ; so that 10lb. falling 10ft. will do 100 foot-pounds of work. Similarly, when 10 units of current flow from a potential of 90 to a potential of 80, or through a difference of potential of 10 units, then 100 ergs of work per second have been done by the current. Also one erg of work is done by forcing a unit current through unit resistance because it requires unit difference of potential to do this, and because the amount of work that is expended in doing this—namely, one erg—is evidently equal to the amount of work—one erg—done by the current. This absolute unit of potential difference is far too small for practical purposes, hence the practical unit of potential difference, or E.M.F.—namely, the volt—is fixed equal to 10^8, or 100,000,000 times the absolute unit.

The practical unit of current is called the ampere, after Ampère, a French physicist, and is that current which one volt will force through a resistance of one ohm. The absolute unit of current has a value of 10 C.G.S. units, and is too large, so the practical unit, the ampere, is equivalent to one-tenth of the absolute unit, and is generally denoted by the letter C.

The practical unit of resistance, denoted by the letter R, is named the ohm, after Ohm, the famous German physicist, who discovered the elementary law connecting pressure, current, and resistance, and is that resistance which when acted upon by a difference of potential of one volt will permit a current of one ampere to flow through it. The absolute unit of resistance is too small, so the practical unit, the ohm, is made equal to 10^9, or 1,000,000,000 times the absolute unit, and according to the Paris Congress of 1884 the ohm is the resistance of a column of pure mercury, one square millimetre (or ·0016 square inch) in sectional area, and 106 centimetres (or 41·73in.) in length, measured at the temperature of melting ice.

Ohm's Law.

The amount of current which will flow in an electrical circuit depends on the E.M.F., which urges the flow, and on the resistance of the material forming the circuit, which tends to obstruct or oppose this flow.

The flow of current is in proportion to the E.M.F., therefore the greater the E.M.F. the greater the flow; it is also in inverse proportion to the resistance, so that the greater the resistance the less will be the flow of current; hence we have

$$\text{Current} \propto \text{E.M.F.}$$

$$\text{Current} \propto \frac{1}{\text{Resistance}}.$$

Combining these two facts together, it is evident that

$$\text{Current} \propto \frac{\text{E.M.F.}}{\text{Resistance}};$$

hence we may say that the current flowing in a circuit is obtained by dividing the E.M.F. by the resistance,

$$\text{or,} \qquad \text{Current} = \frac{\text{Electromotive Force}}{\text{Resistance}},$$

$$\text{or,} \qquad \text{Amperes} = \frac{\text{Volts}}{\text{Ohms}}, \quad \text{or } C = \frac{E}{R}.$$

By multiplying the resistance of a circuit by the current, we obtain the E.M.F. necessary to force that current through that resistance, or E.M.F. = current × resistance, or volts = amperes × ohms, or E = C × R. By dividing the E.M.F. by the strength of current flowing in the circuit we obtain the resistance of the circuit, or, $\qquad$ Resistance $= \dfrac{\text{Electromotive Force}}{\text{Current}}$;

or, $\qquad$ Ohms $= \dfrac{\text{Volts}}{\text{Amperes}}$, or $R = \dfrac{E}{C}$.

The above is known as Ohm's law, and it is the fundamental law of electricity, since it establishes the relation between E.M.F and current and resistance.

Currents are divided into several kinds according to the construction of the electric machine which generates them, such as continuous currents, alternating currents, pulsating currents, multiphase currents, etc.—each of these will be treated with and explained later on. For the present it suffices to say that the first kind—namely, continuous currents—are known as steady currents, because the effective E.M.F. producing them is practically constant in value, but the rest are unsteady, because the E.M.F. is constantly varying or falling and rising in value. Ohm's law, in its above simple form, is absolutely true for all steady currents, and we have every reason to believe that it is true for all unsteady currents as well, but when applied to unsteady currents, say, for example, to an alternating current, then the flow of current in amperes cannot be found by simply dividing the volts of pressure by the pure ohmic resistance of the wire carrying the current, because the effect of the pressure constantly varying is to produce currents of a secondary and tertiary nature arising from self-induction and mutual induction, etc. In addition to this, an alternating current flows first in one direction and then in the opposite, alternately, a great number of times per second. This brings in more complications, because the flow of current offers a certain opposition to change of direction, which is named the inertia of the current, in comparison with the inertia in mechanics. All these disturbing elements must be recognised, and, if possible, be subjected to calculation. It need scarcely be said that when this is done the complicated formula that would be presented would probably be such that Ohm's law in its simple form would be almost lost to sight; so that when Ohm's law is applied, it must be understood that it is only used for steady currents, except when otherwise stated. For those who are interested in this matter, see the *Electrical Review* of May 20, 1892, page 626.

Calculation of Resistance.

The resistance of a conducting wire depends on three things : (1) Its length. (2) Its diameter, or cross-sectional area. (3) Its specific resistance, or substance it is made of. The resistance of a conducting path varies directly with its length, so that a conducting wire 200 yards long has exactly double the resistance of one that is 100 yards long. The resistance of a conducting path varies inversely as its cross-sectional area, or inversely as the square of its diameter (area being proportional to the square of the diameter), so that the thinner a conducting wire becomes, the greater is its resistance. A wire having an area of one square inch has one-half the resistance of a wire having an area of half a square inch, and if the size of the wires be expressed in their diameters, then a wire of $\frac{1}{2}$in. diameter has four times the resistance of a wire of 1in. diameter, and a wire of 2in. diameter has a resistance of one-fourth of the resistance of the 1in. wire. The square of 1 is $1 \times 1 = 1 = d^2$, the square of $\frac{1}{2} = \frac{1}{2} \times \frac{1}{2} = \frac{1}{4} = d^2$, and the square of $2 = 2 \times 2 = 4$, and since the resistance varies inversely as the square of the diameter, we therefore have—

First wire $d = 1$in. $\therefore d^2 = 1 \therefore \dfrac{1}{d^2} = 1 \therefore$ ratio $= 1$

Second wire $d = \frac{1}{2}$in. $\therefore d^2 = \frac{1}{4}$ $\therefore \frac{1}{d^2} = 4$ $\therefore$ ratio $= 4$

Third " $d = 2$in. $\therefore d^2 = 4$ $\therefore \frac{1}{d^2} = \frac{1}{4}$ $\therefore$ " $= \frac{1}{4}$.

The resistance of a conducting path varies directly with its " Specific resistance."

The specific resistance of a substance signifies the resistance it has compared with the resistance of some substance taken as a standard, both substances being of absolutely the same dimensions and under the same conditions, such as temperature, etc. The specified dimensions are expressed either in French or in English measures. When the former are used the specific resistance of a substance is that given by a mass having a cross-sectional area of one centimetre and a length of one centimetre ; when the latter is used it is that given by a mass having a cross-sectional area of one square inch and a length of 1in., the resistance in each case being measured lengthwise. The " Relative resistance " of a substance signifies simply the resistance as compared with a standard, no matter what the dimensions are, provided they are the same for each. Respecting the relative resistances of three substances—silver, copper, and iron—it will be found that giving unit resistance to silver, the values for copper and iron will be 1·06 and 6·4, or putting silver at 100, the others will be 106 and 640. From this it is seen, that having two wires, one of copper and the other of iron, both having the same diameter and length, the resistance offered by iron wire will be six times as much as that offered by the copper wire, because $6·4 \div 1·06$ or $640 \div 106 = 6$, roughly.

The specific resistance of pure annealed silver in English measure—that is, of a mass having one square inch area and 1in. length—is about ·000000633, or $\frac{1}{1580000}$ of an ohm, therefore the resistance of a bar of silver of 1in. area and 1ft. long will be ·0000076, or $\frac{1}{132000}$ of an ohm; this is with a temperature of, say, 18deg. C., or nearly 65deg. F., which may be taken as a fair average temperature of the atmosphere in England. Knowing the relative resistance of any substance at the same temperature, its resistance could be easily calculated out per square inch per foot by multiplying the resistance of a similar bar of silver by the relative resistance ; the question of length and area can then be treated according to the laws previously given. The enormous high cost of silver naturally entirely prohibits its use as a conducting wire, and the next best conductor is copper, which is only a little way behind silver in question of conductivity. Although far below silver in price, copper is an expensive metal, ranging from £40 to £50 per ton, according to the state of the copper market. The capital that must be sunk in copper for feeders and distributing mains for a central electric light station comes to a very heavy sum, sometimes one-third of the total cost, and, as may be expected, the price of the metal is bound to become more expensive as time goes on, on account of the increasing annual consumption, due to a great degree to the progress of electric work. This is assuming that no large sources are discovered.

Copper is the only metal that is used for conveying electric currents for lighting purposes, and it is used in nearly all cases for electric transmission of power and traction. There are some tram lines where silicon bronze wire instead of copper is used, but these form an exception. We will now proceed to work out a practical formula for calculating quickly and easily the resistance of a copper conductor.

The resistance of a bar of pure hard-drawn copper 1ft. long and 1in. diameter may be taken as ·0000105, or $\frac{1}{95238}$ of an ohm. The commercial conductivity of good copper should not be under 98 per cent. of that of the pure copper, and since conductivity is the reciprocal of resistance, therefore the resistance of commercial copper should not be greater than 102 per cent. that of the pure copper. Making this correction, we find the resistance of the above bar for commercial copper will now be ·0000107, or $\frac{1}{93459}$ of an ohm. This is for a temperature of 18deg. C.

Three thousand feet, or 1,000 yards, of No. 12 legal standard gauge (L.S.G.) copper wire, having a diameter of ·104in. may be taken as measuring about three ohms of resistance at 18deg. C. or 65deg. F., the conductivity being 98 per cent.

To calculate the resistance of any sized copper conductor having conductivity of 98 per cent., apply the following rule : Divide the length in feet by 93,000 times the square of the diameter in inches ; the quotient will give the resistance in ohms. That is,

$$\text{Ohms} = \frac{\text{Length in feet}}{(\text{diameter in inches})^2 \times 93,000} ;$$

or,

$$R = \frac{L}{D^2 \times 93,000}.$$

When resistances are very high they are expressed as megohms, one megohm signifying one million (1,000,000) ohms.

When resistances are very low, they are expressed as microhms, one microhm signifying the one-millionth part ($\frac{1}{1000000}$) of an ohm.

When currents are very small in value, they are expressed as milliamperes, one milliampere signifying one-thousandth ($\frac{1}{1000}$) part of an ampere, but this small unit is rarely used.

A few examples will now be given of applying Ohm's law, and the rules given for calculating resistances.

1. A dynamo working at 110 volts pressure at its

terminals, feeds a bank of incandescent lamps having a total resistance of ·2 of an ohm ; how much current do they take ?

Applying Ohm's law we have $C = \dfrac{E}{R}$ and inserting values for E and R we get $C = \dfrac{110}{·2} = 550$, hence 550 amperes is the current taken by the lamps.

2. How many volts are required to force a current of 48 amperes through a circuit having a resistance of three ohms ?

$$E = C R = 48 \times 3 = 144.$$

Hence 144 volts would be required.

3. Find the resistance of a circuit where 40 incandescent lamps in parallel, each taking ·5 of an ampere, are run at a pressure of 100 volts.

The total current will be $40 \times ·5 = 20$ amperes, therefore

$$R = \frac{E}{C} = \frac{100}{20} = 5.$$

Hence the resistance is five ohms.

4. A certain copper wire has a resistance of three ohms. If this were replaced by an iron wire twice as long and half the diameter, what would be its resistance ?

First of all, the relative resistance of iron is, say, 6·4, and that of copper is 1·06, hence the iron wire has $(6·4 \div 1·06)$ times as much resistance as the copper wire,

or, $$\frac{6·4}{1·06} \times 3 = \frac{19·2}{1·06} \text{ ohms.}$$

Second, the iron wire has double the length of the copper wire, and has one-half the diameter, so the resistance must now be multiplied by 2 and again by 4,

·or, $\qquad \dfrac{19\cdot9}{1\cdot06} \times \dfrac{2}{1} \times \dfrac{4}{1} = 145$ nearly.

Hence its resistance is 145 ohms.

5. Calculate the resistance of a copper wire 150 yards long and $\frac{1}{4}$in. diameter. Applying our rule we have—

$$R = \dfrac{L}{D^2 \times 93,000} = \dfrac{450}{\cdot0625 \times 93,000} = \cdot077.$$

Hence the resistance is ·077 of an ohm.

6. A dynamo working with 100 volts pressure at its terminals delivers a current of 10 amperes to a group of lamps 200 yards distant; there must not be a greater drop of potential along the line than 2 per cent. What must be the diameter of the wire used?

As the total loss is 2 per cent., this is two volts drop, or one volt drop in sending the current from the dynamo to the lamps, and one volt drop in returning from the lamps to the dynamo. We have now to find what resistance will absorb one volt in forcing 10 amperes through it. This evidently is $\dfrac{\text{volts}}{\text{amperes}} =$ ohms, or $\dfrac{1}{10} = \cdot1$ of an ohm, so that a resistance of one-tenth of an ohm will cause a drop of one volt in sending 10 amperes through this resistance. From this, it is seen that one length of wire between the dynamo and the lamps—namely, 200 yards—must not have a resistance more than ·1 of an ohm; similarly for the second or return wire. The length and resistance are given, and the diameter is wanted.

Now, $\quad R = \dfrac{L}{D^2 \times 93,000} \quad \therefore D = \sqrt{\left(\dfrac{L}{R \times 93,000}\right)};$

$$\therefore D = \sqrt{\left(\dfrac{600}{\cdot1 \times 93,000}\right)} = \sqrt{(\cdot0645)} = \cdot254.$$

Hence the required diameter of the wire is a little over $\frac{1}{4}$in.

I

Conductors and Insulators.

All substances, so far as we know, will conduct electricity to a greater or less extent, and the property which a substance has for conducting, or acting as a medium for the passage of electricity, is named its " conductivity." A good conductor offers little resistance, and a bad conductor offers great resistance ; a bad conductor is called an insulator, so that a good conductor acts as a bad insulator, and a good insulator acts as a bad conductor. The various metals conduct best, hence their conductivity is very high ; silver takes the first place, and is put down as the standard. Copper follows close behind, and the other metals take their place in order. Acidulated liquids come next, then other liquids, such as sea and fresh water, etc. (oils excepted) ; wet wood is considered as a partial conductor, hence great care should be exercised in seeing that the wood enclosing or supporting conductors is sound and dry. Dry wood belongs to insulators ; nearly all all other materials are classed as insulators.

There is no defining line between conductors and insulators, as may be judged by the note just given respecting wood. It is stated that there is no substance that absolutely checks the passage of electricity, and, on the other hand, there is no substance that does not offer *some* resistance. Hence, we may divide substances into two classes—one class to conduct electricity, called " conductors," and the other class to stop electricity, called " insulators."

In the following table the best conductor is placed first. Half-way through the list the conductors have such bad conductivity that the next commences the insulator class, and those following have been put as near as possible in their proper order ; the nearer the end of the list, the worse the conductivity or the better the insulator. It is difficult to say exactly as to the relative position which several insulators at the bottom of the list should occupy. There seems to be little reliable information. Some insulators lose a good deal of their insulating quality as time goes on, so that although tests will give a very high initial result, it is found that they drop off considerably after use, and so become of lower insulating quality than others that were initially lower, but remained fairly constant.

TABULATION 19.

Classed as Conductors.	Classed as Insulators.
Silver.	Dry wood.
Copper.	Cotton.
Silicon bronze.	Marble.
Gold.	Paper.
Aluminium.	Oils.
Zinc.	Porcelain.
Phosphor Bronze.	Wool.
Platinum.	Silk.
Wrought iron.	Sulphur.
Tin.	Sealing-wax.
Cast steel.	Resin.
Lead.	Vulcanised bitumen.
German silver.	Mica.
Platinoid.	Guttapercha.
Mercury.	Flint glass.
Carbon.	Shellac.
Acids.	Vulcanised indiarubber.
Sea-water.	Ebonite.
Fresh water.	Paraffin wax.
The body.	Dry air.
Wet wood.	

The various woods possess a great difference in resistance; the best insulator is put first in the following list:

Teak (best).	Lignum Vitæ.	Pine.
Walnut.	Rosewood.	Mahogany (worst).

Tabulation 20 gives the relative conductivity of some of the more important materials used as conductors and insulators.

The last seven materials are insulators of the best kind,

and their enormous resistance as compared with the conducting metals can be observed.

TABULATION 20.

Material.	Conductivity.	Material.	Conductivity.
Silver (annealed)..	100	Mercury	1·6
Copper (hard drawn)	92	Carbon	0·05
Silicon bronze......	90	—	—
Phosphor bronze..	24	Mica	$\dfrac{1}{6 \times 10^{17}}$
Platinum............	16·6	Guttapercha	$\dfrac{1}{30 \times 10^{17}}$
Wrought iron......	15·6	Flint glass	$\dfrac{1}{300 \times 10^{17}}$
Cast steel	9·6	Shellac	$\dfrac{1}{600 \times 10^{17}}$
Tin	11·4	Indiarubber (vulcanised)	$\dfrac{1}{1,000 \times 10^{17}}$
German silver ..	7·2	Ebonite	$\dfrac{1}{1,900 \times 10^{17}}$
Platinoid..	4·4	Paraffin wax	$\dfrac{1}{2,300 \times 10^{17}}$

It may be mentioned that 10^{17} signifies one hundred thousand million millions (100,000,000,000,000,000).

Heating Effects of the Current.

The resistance of all metals rises in proportion to the rise of temperature, the only exception being carbon, the resistance of which falls in proportion to its rise of temperature. For copper the increase of resistance is about one-fifth of 1 per cent. for every degree F., or about three-eighths of 1 per cent. for every degree C.; suppose a copper conductor at an atmospheric temperature of 65deg. F. has a resistance of one ohm, then at a temperature of 95deg. F. its resistance would be $1 + 1$ $(30 \times \cdot002)$ $= 1 + \cdot06 = 1\cdot06$ ohms. With guttapercha, the resistance falls as its temperature rises, and this is the case with liquids. It was stated that there is no material that offers no resistance to a flow of electricity, and it was also stated that energy was spent when a current was forced along a conducting path against the resistance of the path. This energy is wasted on the conductor, and appears in the form of heat, the heat energy being exactly equal to the amount of electric energy that produced it. When a unit of current is forced through a unit resistance, one erg of work is done, and to do this requires one unit of difference of potential. When a mass falls through a difference of level the work done is measured by the weight of the mass multiplied into the difference of level or distance through which it falls ; similarly with electricity, the difference of potential corresponds to the distance, and the quantity of electricity corresponds to the weight of the mass : hence difference of potential multiplied by the current will likewise measure the electric work done. One ampere flowing for the space of one second under an electrical pressure of one volt will therefore do $\cdot1 \times 10^8$ $= 10^7 = 10,000,000$ ergs of work per second, because one ampere $= \cdot1$ of a C.G.S. unit, and one volt $= 10^8$ C.G.S. units, 10^7 ergs per second $=$ one watt, hence one watt is the practical unit of electric energy. So that we have

$$\text{amperes} \times \text{volts} = \text{watts} ;$$

746 watts make 1 h.p., therefore we may measure electrical power by multiplying the pressure in volts by the current in amperes, and dividing by 746, or

$$\frac{\text{volts} \times \text{amperes}}{746} = \text{horse-power.}$$

The output, or capacity, of dynamos is usually reckoned

by kilowatts—a kilowatt signifying 1,000 watts. Thus a dynamo designed to work at 110 volts pressure, and to give a maximum current of 500 amperes, is said to have an output of $\dfrac{110 \times 500}{1,000} = 55$ kilowatts ; now, since 746 watts equals 1 h.p., therefore one kilowatt is, roughly, equal to $1\frac{1}{3}$ h.p., or 1 h.p. is equal to three-quarters of a kilowatt.

In the same way as steam power is sold or measured by horse-power hours, so electrical power is sold or measured by kilowatt-hours—one kilowatt-hour being named a Board of Trade Supply Unit. One kilowatt-hour, or, briefly, one "supply unit," is defined by the Board of Trade to signify 1,000 watts of electrical energy acting for the space of one hour. Hence one "supply unit" = 1,000 amperes flowing under a pressure of one volt, or 100 amperes under a pressure of 10 volts, or 10 amperes under a pressure of 1,000 volts, or generally 1,000 volt-amperes, for the space of one hour.

The amount of electric energy that is spent or wasted in the conducting wires is evidently found by multiplying the current that is flowing by the difference of potential that there is between the ends of the conductor, for this difference of potential measures what is technically called "drop" of potential, and signifies the pressure that is absorbed or lost in forcing the current through the resist- ance of the length of conductor. According to Ohm's law, volts = amperes × ohms, or $E = C R$, and energy wasted in the conductor is "drop" in volts × amperes, or $E C$; therefore, substituting the value of the volts, or E, we have energy wasted = $(amperes)^2 \times ohms$, or $C^2 R$. Hence the energy wasted in the form of heat in a conductor is obtained by multiplying the square of the current by the resistance of the conductor.

The amount of heat energy generated by the flowing of the current through a conductor can be easily calculated. By inserting a coil of German-silver wire of known resistance in a bath of paraffin oil, and passing a known current through the wire, the heating effect will be shown by the rise of temperature of the oil. From numerous and careful experiments made, it is found that multiplying the electric energy wasted in the wire by the number ·24 gives the heating effect of the current per second in heat units, or calories; hence we have the following formula :

$$H = C^2 R t \times ·24,$$

where H = calories = number of grammes of water raised 1 deg. C., and t = time in seconds.

It is thus seen that the energy wasted is proportional to the square of the current, and directly to the resistance. The rapidity with which a substance dissipates or throws off heat is named the emissivity, and the heat generated in a conductor should be thrown off as quickly as it is produced there. The emissivity depends on the cooling surface, and this is measured by the circumference of the wire multiplied by its length, and since circumference is proportional to diameter, hence the cooling surface of a wire is proportional directly to diameter. But the current that a wire will carry is proportional to the square of the diameter, since double the diameter gives four times the area, and therefore four times the current can be carried. The resistance is now only one-fourth of what it was before, so that the electrical energy wasted by a wire of double diameter is $C^2 R = 4^2 \times \frac{1}{4} = 16 \times \frac{1}{4} = 4$ times what would be wasted in a wire of unit diameter, each wire having the same current density. We thus see that while the heat energy generated is four-fold, the cooling surface is only twofold; consequently the larger wire must become much more heated than the smaller wire. So the smaller a wire becomes, the greater is the current it will carry in proportion to its area ; and the larger a wire becomes, the less is the current it will carry in proportion to its area. For example, suppose a wire of $\frac{1}{2}$in. diameter will carry 400 amperes without overheating, it will be found that a wire of 1in. diameter carrying its proportional current — namely, four times 400, or 1,600 amperes—would become much hotter than the first, and a wire of $\frac{1}{4}$in. diameter, carrying 100 amperes, would become much cooler. The quantity of current per unit area is the same—that is, the current density, as it is called, is the same for all three wires ; the explanation lies in the fact that the cooling surface is proportionally largest in the thinnest wire, and smallest in the thickest wire. Divide the 1in. circular wire into four separate circular strands or parts, each having one-fourth the sectional area of the solid wire ; the cooling surfaces of the four stranded and one solid conductors will be in the ratio of $\sqrt{4} : 1 = 2 : 1$, because each strand will have one-half inch diameter, and four times one-half equals twice one.

If this same solid conductor be divided into nine strands, each having one-ninth of the total sectional area, then ratio of cooling surface, as compared with the solid conductor, will be $\sqrt{9} : 1 = 3 : 1$; so that the cooling surface in this case will be trebled. The ratio can be at once obtained for any number of strands, putting the solid conductor at unity and taking the square root of the number ($\sqrt{n}$) of the strands for the divided conductor.

We are now in a position to calculate out the temperature to which a given conductor will rise when a given current is sent through it, and by fixing this temperature at a safe limit, which must not be exceeded, we can then get at the greatest current that can flow without raising the temperature of the conductor beyond this fixed safe limit.

A wire which would become heated to about 24deg. C. or 75deg. F. while at atmospheric pressure, would, with the same current flowing through, become red hot when the pressure is reduced to, say, $\dfrac{1}{1\cdot 7 + 10^6}$th part of an atmosphere, thus proving that far more heat is conveyed away by convection than by radiation. The latter pressure is somewhere about that which remains in an incandescent lamp when so much air is exhausted that it may be said practically to be a vacuum.

The emissivity of copper, or the rate at which heat is dissipated from its surface, is ·0003 of a calorie per second for every square centimetre of exposed surface (one calorie being the heat necessary to raise one gramme of pure water 1deg. C.), for every difference of temperature of 1deg. C., between the wire and surrounding bodies. Hence the total heat emitted = ·0003 T S calories per second, where T = difference of temperature in degrees C., and S = exposed cylindrical surface in square centimetres. It has been shown that the total heat generated in the wire is expressed by ·24 C² R calories per second, where C = the current in amperes, R = resistance in ohms ; and since the heat emitted or dissipated should be equal to the heat generated, therefore ·0003 T S = ·24 C² R, or

$$T = \frac{\cdot 24\ C^2\ R}{\cdot 0003\ S} = \frac{800\ C^2\ R}{S}.$$

So that multiplying the electric energy wasted in the wire by 800, and dividing by the square centimetres of exposed surface, will give the number of degrees C. of temperature the wire will be raised above the temperature of the atmosphere. We have also,

$$C = \sqrt{\left(\frac{\cdot 0003\ T\ S}{\cdot 24\ R}\right)}.$$

This formula gives us the maximum current a wire will carry so that its temperature shall not rise more than T degrees C. above that of the atmosphere. In practice, conductors should not be heated more than 50deg. C. or 90deg. F. above the atmosphere; so that we deduce the following formula :

$$C = \sqrt{\left(\frac{\cdot 0003 \times 50 \times S}{\cdot 24 \ R} \right)} = \cdot 25 \sqrt{\frac{S}{R}}.$$

Expressing the exposed surface in square inches, we have

$$C = \cdot 64 \sqrt{\frac{S}{R}}.$$

Current density is expressed as so many amperes per square inch, and we will apply this last approximate rule and see what current density it gives. A wire to have one square inch of cross-sectional area must have a diameter of a little over 1⅛in., and the circumference will be 1·128 × 3·1416 = 3·543in. Assume the wire is 200ft. long, therefore the cylindrical exposed surface will be 2,400 × 3·543 = 8,563 square inches, and the resistance in ohms at 50deg. C. above the atmosphere may be put down, roughly, at ·002. Inserting our values, we have—

$$C = \cdot 64 \sqrt{\frac{8,563}{\cdot 002}} = 1,324 \text{ amperes.}$$

Hence, in order that the wire shall not heat up more than 50deg. C. or 90deg. F. above the temperature of the atmosphere, the current must not be more than about 1,324 amperes for a conductor having one square inch area.

When the same amount of current flows through wires of different diameters, the temperature of the wire is inversely proportional to the third power of the diameter, because halving the diameter of a wire gives—first, four times the resistance, and since the current remains constant, this means four times the heating effect ; second, the mass has now only one-half the cooling surface, therefore we have four times the heat emitted from one-half the cooling surface, so that its temperature will be eight times as much. Expressed in symbols, this law is $T \propto \dfrac{1}{D^3}$, where T equals temperature and D equals diameter.

Simple and Divided Circuits.

The most simple electric circuit possible is shown in Fig. 9 ; in this, as well as in any other circuit, simple or divided, the whole circuit may be classed into three distinct portions—namely, (1) the generator, or source of electricity; (2) the conducting wires with switches, etc.; (3) the source of light, heat, or power.

As a generator, the dynamo is mentioned, and the electricity so produced, after being guided whither it is required, is made to do work by producing a source of light, as an arc or incandescent lamp, by producing heat, as an electric heater, or by producing mechanical power, as an electro-motor.

The first portion of the circuit, that of the dynamo, and limited by T T_1 on either side, is known as the internal circuit, whilst the other two portions, those belonging to the conducting wires, and the lamp, or other apparatus, are known as the external circuit.

In Fig. 9, T signifies the positive terminal of the dynamo, where the electricity is supposed to make its exit or flow into the external circuit ; at T_1 it leaves the external circuit and completes the circuit by entering the generator, whence it came. The distinction between E.M.F. and difference of potential has been fully explained, and here we can now with advantage apply this distinction. The E.M.F., or maximum

difference of potential, has its seat in the coils of the armature
of the dynamo, and is that which urges the flow of current
through the whole of the circuit, external and internal, so
that the current starting from one point of the armature
coils enters the external circuit by the terminal T, flows along
the conducting wire, through the lamp, back through the
other conducting wire, and so by terminal T₁ into the armature coil whence it started, thus completing the entire

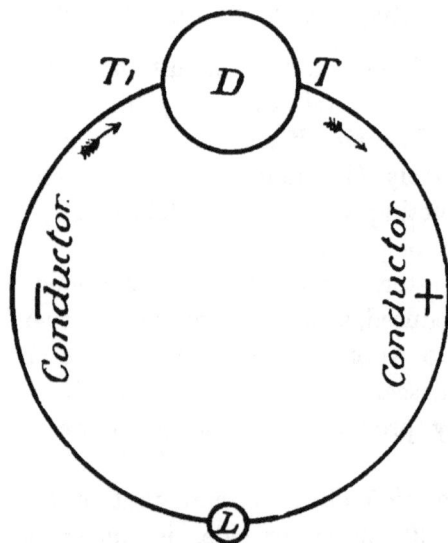

FIG. 9.

circuit.   It is evident that a certain amount of potential will
be lost by forcing the current from its starting point to the
terminal T, because the current has to be urged through the
resistance offered by the wire comprising the armature coils
and the field-magnet coils.   Again, a certain amount of
potential will be lost by forcing the current from the
terminal T₁ back to its starting point, again due to the
resistance offered by the armature coils and the field-magnet

coils. Multiplying the internal resistance, which is that between the two terminals, T and T_1, by the square of the current flowing, or what amounts to the same thing, multiplying the drop of potential between T and T_1 by the current, will give the electric energy that is lost in the dynamo. The second portion of the circuit comprises the two conducting wires, one leading to the lamp, called the positive or + wire, and the other leading from the lamp, called the negative or — wire. Both may be taken as similar; therefore, multiplying the resistance of the two wires by the square of the current will give the electric energy lost in the two conductors.

The third, or remaining portion, is the lamp, or other apparatus, and the energy utilised here is obtained in a similar way. Adding the three results together gives the total energy generated by the dynamo, and since the energy that is sent into the external circuit (through the conductors and lamps) is the amount given out by the machine, hence the ratio between the total, or internal + external power, and the external power gives the efficiency of the machine. To give an idea of the various losses, it may be said that 10 per cent. of the total electric power developed is lost in the internal or machine part of the circuit, 5 per cent. in the conducting wires, and the rest utilised in the lamps. With the same amount of current in all parts of the circuit, the loss will be proportional to the drop of volts. Using these values, and having a working difference of potential between the terminals of the dynamo of 100 volts, the maximum potential in the internal circuit may be about 112; upon reaching the positive terminal, the potential there will be 106, six being lost. Allowing $2\frac{1}{2}$ per cent. for the positive conductor, we have about 103 at the point of the circuit where the current enters the lamp. On passing through the lamp the resistance is very high, and it is here that the power is utilised, so that on coming out of the lamp the potential is very low, say, only nine. Allowing an equal drop along the negative conductor, we have a potential of six at the negative terminal of the machine, the remaining six being lost in the internal circuit.

We have discussed the resistance of a circuit when there is only one path for the current to flow along, but what effect upon the current would divided paths have ? This is a matter that may lead to complicated calculations where the divided paths are numerous and of irregular resistance, etc. ;

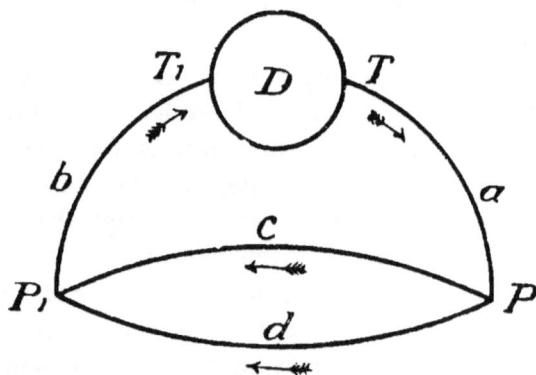

FIG. 10.

so we will begin with the simplest and gradually work towards those not so simple, and when a certain limit has been reached more advanced problems will merely be referred to, since they would not be suitable here.

Study Fig. 10. At the point P of the conductor a it will be seen that the conductor divides into two paths, c and d ; let each have the same length and the same diameter, this diameter being the same as that of a and b ; at the point P_1, c and d join together on to the conductor b. Resistance is inversely proportional to cross-sectional area, and the combined area of c and d must be double that of either of them alone ;

hence the combined resistance must be one-half of either of them alone. The amount of current in a circuit is the same at all points, so that whatever current flows past the point P of a into the two divided paths, c and d, the same amount must arrive at the point P_1 of b. The area of c is equal to the area of d, hence the current will divide equally at P, one half going through c, the other half going through d; at P_1 the two halves unite again. Putting the resistance of c and d at 10 ohms each, the combined resistance will be $10 \div 2 = 5$ ohms, and if 50 amperes flow past P, 25 will flow past c, and 25 past d. If a and b were divided into four paths, instead of two, all four having the same diameter and the same length, consequently the same resistance, then the combined resistance of the four would be one-fourth of that offered by a single one; hence the current would split into four paths, and for any number of paths having equal resistance, the combined resistance is obtained by dividing the resistance of a single path by the number of paths there are, or,

$$ R = \frac{r}{n}, $$

where R = the combined or total resistance ;

r = the resistance of a single path ; and

n = the number of paths.

When two or more paths are open to a current these paths are said to be in "parallel," so that the two wires c and d in Fig. 10 are in parallel. When the end of one path is joined on to the end of another path so as to make one continuous length, then these two paths are said to be in "series," so that the single path a is in series with the divided paths, c and d; also c and d are in series with the single path b, hence a, c and d, b are all in series, and the total resistance of these three paths would be resistance of a + combined

resistance of c and d + resistance of b.   Let $a = 20$ ohms
$c = 5$ ohms, $d = 5$ ohms, and $b = 20$ ohms.   The first thing
to be done is to find the equivalent or combined resistance of
c and d; this is, $5 \div 2 = 2 \cdot 5$, so that the combined resistance of the two paths c and d is the same as if they were
replaced by a single path having a resistance of $2 \cdot 5$ ohms.
All being in series, the total resistance is $20 + 2 \cdot 5 + 20$
$= 42 \cdot 5$ ohms.   In Fig. 11 the calculations are a little more
difficult.   We have again a divided path between the two
points P and P_1, but the two conductors c and d have in

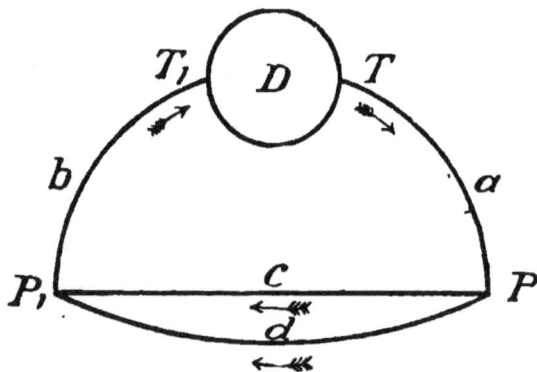

FIG. 11.

this case unequal resistances, due, say, to length, therefore c
is drawn shorter than d.   We will first find the combined
resistance and then the way the current divides itself.   When
treating of unequal resistances, it is convenient to convert
them into conductivities, and having found the value of the
combined conductivity of all the paths, to then reconvert
them into combined resistance.   Conductivity of a path is in
inverse proportion to the resistance, and resistance is in
inverse proportion to the conductivity—that is, putting k
for conductivity and r for resistance, we have $k \propto \dfrac{1}{r}$ and

$r \propto \dfrac{1}{k}$; so that the conductivity is the reciprocal of the resistance, and *vice versâ*. Instead of considering two resistance paths we will consider two conducting paths, one of which conducts better than the other because it has less resistance than the other. It is evident that the combined conductivity of the two conducting paths is equal to their sum : giving a resistance of r_1 to c, and r_2 to d, their respective conductivities will be $\dfrac{1}{r_1}$ and $\dfrac{1}{r_2}$.

Hence,

$$\text{Combined conductivity} = \frac{1}{r_1} + \frac{1}{r_2},$$

and putting R for the combined resistance, we have $\dfrac{1}{R}$ for the combined conductivity ;

therefore
$$\frac{1}{R} = \frac{1}{r_1} + \frac{1}{r_2} = \frac{r_2 + r_1}{r_1\, r_2},$$

and converting conductivities into resistances, we get

$$R = \frac{r_1\, r_2}{r_1 + r_2}.$$

Putting the above law into words, this means that the combined resistance of two unequal resistances in parallel is obtained by dividing their product by their sum ; for let the resistance of c be 15 ohms, and that of d be 20 ohms, then the combined resistance will be

$$\frac{15 \times 20}{15 + 20} = \frac{300}{35} = 8\cdot 57 \text{ ohms.}$$

The preceding rule only applies to two resistances ; beyond two, the rule is more complicated. Take a case of four unequal resistances, say, $r_1, r_2, r_3\ r_4$, we have

K

$$\frac{1}{R} = \frac{1}{r_1} + \frac{1}{r_2} + \frac{1}{r_3} + \frac{1}{r_4}$$

$$= \frac{(r_2 r_3 r_4) + (r_1 r_3 r_4) + (r_1 r_2 r_4) + (r_1 r_2 r_3)}{r_1 r_2 r_3 r_4}.$$

Therefore, $R = \dfrac{r_1 r_2 r_3 r_4}{(r_2 r_3 r_4) + (r_1 r_3 r_4) + (r_1 r_2 r_4) + (r_1 r_2 r_3)}$;

this signifies that the combined resistance is in the form of a fraction whose numerator is the product of all the resistances, and whose denominator consists of as many expressions as there are resistances, each expression being obtained by dividing the product or the numerator by a different resistance. Putting $r_1 = 5$, $r_2 = 6$, $r_3 = 7$, $r_4 = 8$, we shall obtain a combined or equivalent resistance of

$$R = \frac{5 \times 6 \times 7 \times 8}{(6 \times 7 \times 8) + (5 \times 7 \times 8) + (5 \times 6 \times 8) + (5 \times 6 \times 7)}$$

$$= \frac{1,680}{1,066} = 1\cdot57 \text{ ohms.}$$

Coming to the matter of division of current in parallel paths of unequal resistance, the current divides itself according to the conductivity of the path, or inversely to the resistance. In the example worked out, where the resistance of c is 15 ohms, and that of d is 20 ohms, let the current be 35 amperes at the point P. The conductivity of $c = \frac{1}{15}$, that of $d = \frac{1}{20}$. The whole of the current, or 35 amperes, passes through the two paths, c and d, and hence through the combined conductivity, $\frac{1}{15} + \frac{1}{20}$. The current divides according to the conductivity; therefore, as the whole current is to the combined conductivity, so the current in c will be to the conductivity of c.

Hence current in $c = 35 \times \frac{1}{15} \div (\frac{1}{15} + \frac{1}{20}) = 20$ amperes.

,,   ,, $d = 35 \times \frac{1}{20} \div (\frac{1}{15} + \frac{1}{20}) = 15$ amperes.

Where the number of paths in parallel are more than two, the current for any particular path is obtained by dividing the total current by the combined conductivity of all the paths, and then multiplying by the conductivity of the path in question.

Fig. 12 illustrates a circuit having some resistances in parallel and some in series. Let $a = 10$ ohms, $b = 10$ ohms $c = 6$ ohms, $d = 7$ ohms, $e = 5$ ohms, and $f = 5$ ohms.

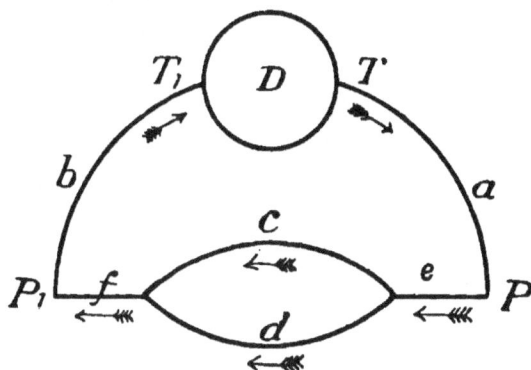

FIG. 12.

It is necessary first to eliminate out the combined resistance of the divided path, which will be

$$\frac{6 \times 7}{6 + 7} = \frac{42}{13} = 3 \cdot 23 \text{ ohms.}$$

Substituting this value for the divided paths, c and d, we have now got all the resistances in series, and these have simply to be added up for the total resistance between terminals T and T_1; hence the total resistance $= 10 + 5 + 3 \cdot 23 + 5 + 10 = 33 \cdot 23$ ohms.

Magnets and Magnetism.

The fundamental principle upon which may be based the action of all dynamo-electric machinery was stated in the opening paragraphs; for convenience these remarks are repeated, which are to the effect that "if a length of wire or metal conductor, whose two ends are joined so as to make a complete circuit, be caused to move across the magnetic field so as to cut through, so to speak, the 'lines of force,' then a current of electricity will be generated in that conductor, and will last so long as the movement of the conductor lasts in the magnetic field. The same effect will be produced if the wire is stationary and the magnet be caused to move." Hence the presence of magnetism is necessary in those machines in order to produce an electric current, so before entering upon the province of the dynamo some elementary information concerning magnetism and its laws will be given.

Magnets may be classed in two divisions : (1) natural magnets ; (2) artificial magnets.

The earliest knowledge we have respecting the curious phenomenon of magnetic attraction, is probably that regarding the properties of a certain earthy stone. This peculiar stone possesses the property of attracting iron, and was named a "lodestone" or "magnet," its attractive virtue being termed "magnetism." The lodestone is a natural magnet, and the earth is a huge natural magnet, the true or magnetic north pole being to the west of the geographical north pole, and the true or magnetic south pole being to the east of the geographical south pole.

Earth-currents seem to be the cause of the magnetic storms and disturbances that take place around the earth. As an example of the effect of unequal or different electrical conditions existing between the atmosphere and the earth, the aurora borealis may be mentioned.

Artificial magnets may be divided into two kinds : (a) permanent magnets ; (b) electromagnets.

If a piece of steel be rubbed by a piece of lodestone, the

steel is at once endowed with magnetism, and the harder the steel is the longer it takes to impart magnetism; but against this can be put a longer period of retention of magnetism. These magnets are named "permanent magnets," because they retain their magnetism for years. Soft iron cannot be made into a permanent magnet, although after being subjected to magnetic influences it will always retain a feeble and insignificant amount of magnetism too weak to have much attractive power. Electromagnets are made by passing a current of electricity a number of times around a bar of soft iron, so that the bar must be wound with an insulated wire. When the current is sent through these coils the bar is magnetised, but when the current is stopped the bar loses all its magnetism. Some of the early dynamos were built with permanent steel magnets; now, only electromagnets of wrought or cast iron are used.

When a current of electricity flows through a conductor, it is encircled by "magnetic whirls," or "lines of force," as shown in Fig. 13 and also in Fig. 14. These whirls are seen to be at right angles to the length of the conductor; hence, at right angles, or perpendicular to the flow of current in the conductor. If the conductor be coiled into a circle, this circle will embrace a great number of short lengths or parts of curved lines forming the magnetic whirls; and since the magnetic whirls become less dense the more they are distant from the conducting wire, therefore the lines that pass through the central parts of the circular coil will be less dense than those that pass through close by the periphery of the circular coil. Let the conductor be coiled twice round, the respective magnetic whirls of each turn would merge together in some parts and neutralise each other in other parts, and so form one larger whirl or loop of force, surrounding both the turns. Increase the turns to a considerable number, by winding the conductor in the form of a helix, or spiral, the turns being close together, and parallel with each other as much as possible, coincidental whirls will then merge together and form one large whirl or line of force, which will surround the helix, as shown by any of the dotted lines in Fig. 15, where, owing to the length of the helix, the lines of force map out practically a straight path

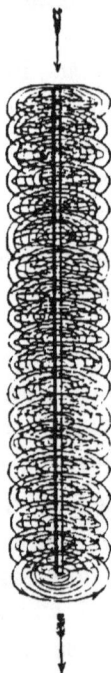

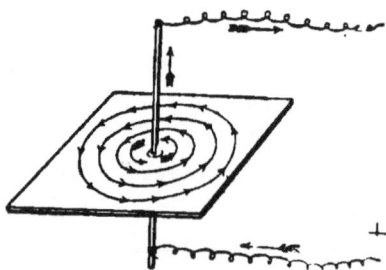

FIG. 13.   FIG. 14

along the length of the helix. We have thus the whole of the hollow space inside the helix threaded by lines of force, which curve round at one end, flow along the outside of the helix and enter again at the other end, thus forming a complete circuit. The direction in which these lines of force may be said to flow depends on the direction of the current that flows in the coils. When a flow of lines of force is

:spoken of, it must be remembered that the term lines of force is only a figurative expression illustrative of the path along which magnetic force tends to act, so that direction of flow may be taken to signify direction of force. A line of force always flows in a closed circuit, in the same way that a current can only flow when there is a closed circuit ; consequently, every line of force that issues from any part of an electromagnet describes a more or less symmetrical circuit (according to surrounding bodies) whatever extent of space it may have to travel through, and so returns to the point from which we trace its departure.

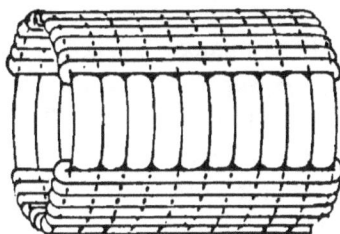

FIG. 15.

Any space through which lines of force pass is termed a magnetic field, therefore a magnetic field surrounds a wire carrying a current, and a magnetic field exists both inside and around a helix of wire. Such a helix through which a current flows is technically called a " solenoid," the insulated wire forming the helix being usually wound on a brass bobbin, resembling somewhat a cotton-reel in shape, the channel formed by the flanges being filled with insulated wire.

The strength or intensity of a magnetic field at any point is denoted by the number of lines of force that thread through one square centimetre ($\cdot$155 square inch) of area of the air, measured at right angles to the direction of flow.

Thus if the internal diameter of a solenoid were 10cm., and there were 4,000 lines passing through, then the density, or the number per square centimetre, would be the total number of lines divided by the area in square centimetres, or 4,000 ÷ 78·54 = 51 nearly. So that the strength of field is 51.

Upon introducing a bar of soft iron just within the hollow of the solenoid, it will be sucked up with force and held there in suspension. This is due to the fact that the mass of iron is brought within the attractive range of the lines of force constituting the magnetic field, and therefore it is attracted; but whilst the iron bar fills up, or partly fills up, the hollow of the solenoid, a great change is made in the magnetic field. This change is that the iron has a far greater conducting power for magnetic lines than the air. Consequently, a far greater number will now flow through the mass of iron than what previously flowed through the space occupied by the air. It is this flow of lines through the iron that converts the iron into a magnet, and so the reason is explained why a piece of iron is magnetised by winding it with an insulated wire through which a current is passed, because putting an iron bar inside the hollow of a solenoid has the same effect as winding the bar with wire. When used as a solenoid, the bobbin carrying the wire is fixed, whilst the iron bar is supported, so as to be perfectly free to move up and down within the solenoid; when used as an electromagnet, the bobbin of wire and the iron core are both fastened together and stationary.

Solenoids are used when a long range of pull is required, say several inches, but this extended pull is only comparatively weak.

Electromagnets are used when a very powerful pull is desired, but this pull can only be exerted over a very short distance, measured in fractions of an inch. In the first case

the mechanism to be moved is fixed to the free iron bar, or
"plunger," as it is called; in the second case, the mechanism
is fixed to an iron mass, called the "armature," placed a very
short distance away from the end or pole of the electromagnet. The electromagnet attracting the armature moves
the mechanism.

Magnetic Properties of Iron.

We have every reason to think that all matter with which
we are acquainted, of whatever kind, is susceptible to magnetism; by the word "susceptible" we wish to imply
"capable of being magnetised." With the exception of
several bodies, all the rest are susceptible to so minute a
degree that it is practically negligible, and hence they may
be termed non-magnetic bodies, in order to distinguish them
from those few which have marked susceptibility, and which
may be termed "magnetic bodies." The non-magnetic
bodies are very nearly equal so far as we can judge, and the
air is taken as unity. Classed among these are a few bodies
with a susceptibility less than air; these are named dia-
magnetic bodies, but the strongest of them, which is
bismuth, has a value of ·999 that of air.

The chief magnetic bodies are iron, nickel, and cobalt;
the two latter are about equal, but iron is at least three
times as susceptible.

Iron may thus be said to stand out alone, by being so
much beyond all other bodies in respect of susceptibility.
Indeed, for the magnetic circuit, and for all apparatus
receiving magnetic induction, iron—whether wrought iron,
cast iron, or steel—occupies the same position as copper does
for the electric circuit. This being so, the phenomenon of
magnetism and its effects on bodies will be confined to iron,
because it is only iron that will be dealt with.

It was explained that when a mass of iron, like an iron
bar, was introduced into a space previously occupied by air

in which a magnetic field existed, as the interior of a solenoid, the number of lines of force were greatly increased, so that for one line per square centimetre through the air, probably 1,000 per square centimetre would thread their way through the iron. The iron appears to have a sort of "multiplying power"—this property of the iron is named its "permeability," a matter of great importance. Permeability may be considered in some degree as an analogue to specific conductivity in the electric circuit, but there is this great difference. Specific conductivity remains constant, neglecting rise and fall of temperature of the wire, irrespective of the amount of current, whereas permeability is not constant for varying number of lines—so inconstant that its values for the same piece of iron may range from unity to several thousand. The permeability of a vacuum is arbitrarily fixed at unity, that of air is so near that it is practically unity also, the same with all the so-called non-magnetic bodies, and their permeability may be regarded as constant whether the number of lines be few or many. Dia-magnetic bodies have permeability less than unity—that is, they repel or turn aside part of the lines of force, so that if a piece of bismuth were placed in a magnetic field it would have a "reducing power," and of the number of lines that would flow through air, ·9998 of the number would flow through bismuth, or for 10,000 that pass through air, 9,998 would flow through bismuth, the remaining two being repelled.

The number of lines of force per square centimetre that flow through air and form a magnetic field are denoted by the lettter H. When iron displaces the air, the increased number of lines per square centimetre that now pass through the iron is named the "magnetic induction," denoted by the letter B. The ratio of B to H indicates the multiplying power of the iron, or its permeability, denoted by the Greek letter μ, hence $\mu = \dfrac{B}{H}$, and $B = \mu H$.

Suppose in an air-space there are four lines per square centimetre, and the air is displaced by soft iron, the lines through the iron will be, say, 9,000, hence the value of μ will be 2,250, and this is the multiplying power of the iron ; or we may say that the iron conducts or receives the magnetism 2,250 times better than air does. Now let there be 200 lines per square centimetre in air, again displaced by iron, we shall only get 18,000 lines; hence μ is now only 90. From this it is evident that the permeability decreases very rapidly in value as the iron becomes more saturated with lines—that is to say, the greater the magnetic induction the less is its multiplying power. When a certain degree of saturation has been reached, it is unprofitable to force more lines of force through the iron, because, owing to the diminished value of μ, an enormous magnetising force would be required.

TABULATION 21.—Annealed Wrought Iron.

H.	B.	μ
1·66	5,000	3,000
4	9,000	2,250
5	10,000	2,000
6·5	11,000	1,692
8·5	12,000	1,412
12	13,000	1,083
17	14,000	823
28·5	15,000	526
50	16,000	320
105	17,000	161
200	18,000	90
350	19,000	54
660	20,000	30

TABULATION 22.—Grey Cast Iron.

H.	B.	μ
5	4,000	800
10	5,000	500
21·5	6,000	279
42	7,100	133
80	8,000	100
127	9,000	71
188	10,000	53
262	11,000	37

In Tabulations 21 and 22 are given figures calculated
from experiments made by Dr. Hopkinson on specimens of
annealed wrought iron and grey cast iron, and these figures
represent fairly the quality of the iron used by dynamo-
makers.   H signifies the magnetising force, or the number of
lines of force per square centimetre that there would be in
air.   B signifies the magnetic induction of the iron, or the
number of lines of force per square centimetre. μ signifies
the permeability, or multiplying power of the iron.

In Fig. 16 these values are plotted into curves, values of B
being plotted horizontally, and those of H vertically ; the
lower curve being for grey cast iron, and the higher curve for
annealed wrought iron.   On referring to Tabulation 21, it is
seen how enormously the permeability is changed, descending
from 3,000 to 30, whilst in Tabulation 22 it drops from 800
to 37.   It is interesting to carefully follow the curve, as it
gives a good idea of how the iron is gradually magnetised.
First, the two values, H and B, increase fairly together,
and this part of the curve is plotted from small values
of H and B, not given in the tabulations ; a little further
on the curve takes a sudden upward direction, almost
perpendicular.   This means that a very little increase of H
gives an enormous increase to B.   The curve now bends
round and then pursues its course in a straight line, still
rising, so that it makes a gentle angle above the horizontal.
Beyond the bend, the value of B increases very little as compared with H, just the opposite to the state of affairs below
the bend.   For example, with B at 19,000, and H at 350, it
requires nearly double the magnetising force in order to send
an additional 1,000 lines through the iron.   The curve may
be divided into three portions—the straight part below the
bend, the bend, and the straight part beyond the bend.   In
the first part, the iron may be looked upon as being unsatu-
rated ; in the third part, as approaching saturation.

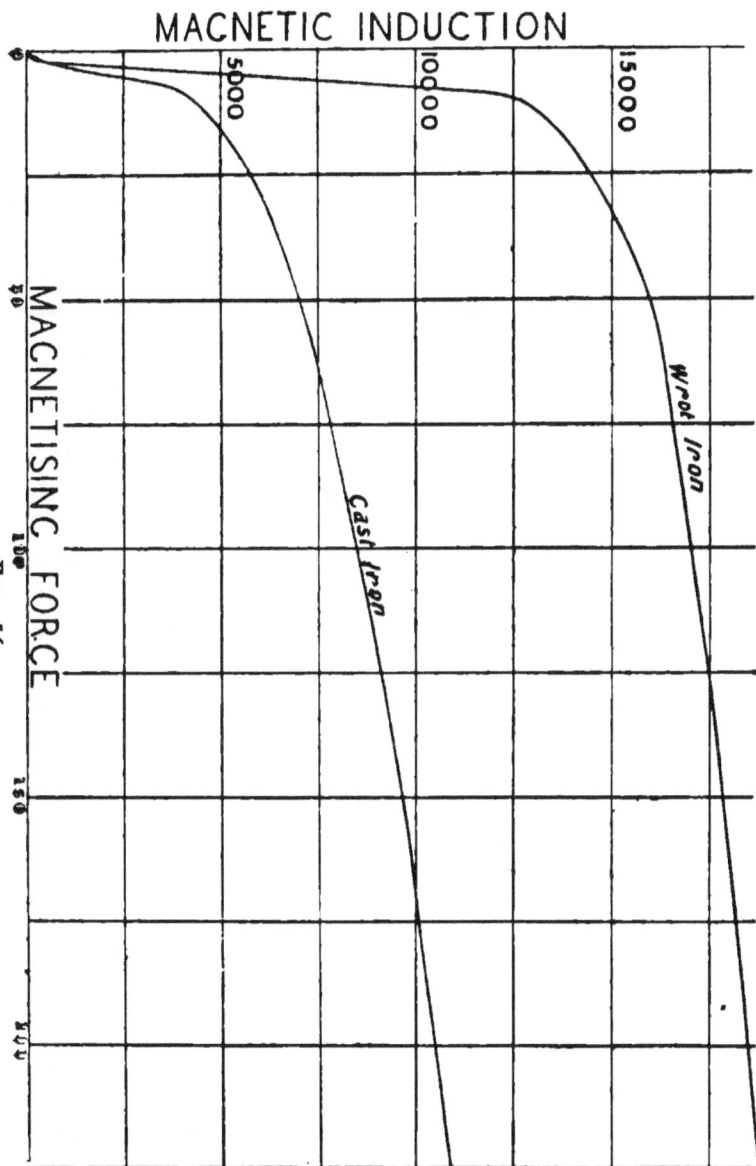

MACNETIC INDUCTION

5000 10000 15000

MACNETISINC FORCE

Wrot Iron

Cast Iron

FIG. 16.

The dia-critical point of saturation signifies that point on the curve where the permeability of the iron is reduced to one-half of its maximum permeability, or when the iron is half saturated. The curve for cast iron is much after the same shape as for wrought iron, but being below iron shows that it is not so susceptible to magnetism, therefore it requires more magnetising force to produce the same induction as iron. The dia-critical point of saturation is about three-quarters that of wrought iron. Dynamo magnets are mostly made of soft wrought iron, and it is usual in dynamo design to allow about 14,000 lines per square centimetre. A greater induction than this would not be economical, because for a small increase there would be the expense of a very much greater magnetising force required, and it would be cheaper to design a larger magnet by using more iron. The last few years cast iron has been used with success. Owing to its lower susceptibility and permeability, a larger mass is necessary. The advantages for it are that it can be cast into any desired shape, and the cost of production is made much less; the economical limit for it is about 10,000 lines. Mild steel has also been adopted, and answers its purpose well; its permeability lies between cast iron and wrought iron, and where the steel is specially prepared it is almost as good as the latter. When the greatest susceptibility is wanted, iron of the very softest kind must be used, and for this, Swedish charcoal iron is the best; next to this comes Lowmoor iron. Great susceptibility is obtained by heating the iron white hot, and then allowing it to cool down gradually, extending the cooling process, say, for several days, by burying it in sawdust, hot ashes, etc.

When a bar of iron that has been magnetised is caused to lose its magnetism by stopping the current in the magnetising coils, it will be found that a certain amount of magnetism is left in the iron; this amount is named the " residual magnetism," and is dependent on the retentiveness of the iron—the softer the iron the greater is its retentiveness. Iron that has been hardened by any process such as hammering, rolling, etc., possesses less retentiveness. The hardness of iron depends on the amount of carbon in its composition. The more carbon there is, the harder becomes the metal, such as hard steel, etc. Mild steel, like wrought iron, has very little carbon; hence, they are soft, possess high permeability and high retentiveness.

No two pieces of iron display the same permeability, and in a number of cases it has been found that two lengths cut off the same bar differ greatly in permeability. It is a matter of the highest importance for a dynamo-builder to become acquainted with the magnetic quality of the iron he buys, otherwise his calculations are thrown wrong. The following high values of induction, B, or lines per square centimetre, have been obtained by Dr. Hopkinson : 18,250 for wrought iron, 19,840 for mild Whitworth steel, 11,000 for cast iron, and 12,408 for malleable cast iron.

Foreign elements in the iron make a very great difference to the permeability, such as tungsten, manganese, sulphur, chromium, aluminium, etc. Upon adding about 12 per cent. of manganese, a compound is obtained which is almost non-magnetic. The temper of the metal has also an effect upon its magnetic qualities. All magnetic bodies, particularly iron, resist any alteration of their magnetic condition; those which are magnetised oppose any attempt to demagnetise them, and those which are not magnetised oppose being magnetised ; and generally, at whatever stage of magnetisation any body may be, there is always a certain opposition shown by the body to any change of its magnetic condition. From this it is very clear that a body which is being continually magnetised and demagnetised, will continually be opposing the continual change of condition.

Such a state of affairs is exemplified by the action of an alternating current upon a mass of iron. The iron is first gradually magnetised, due to the rise of current flowing in one direction ; it is then gradually demagnetised by the falling of the current ; the current again rising, but flowing in the opposite direction, again magnetises the iron, and again falling demagnetises it. These operations are called a magnetic cycle, and are shown by Fig. 17. At the point O

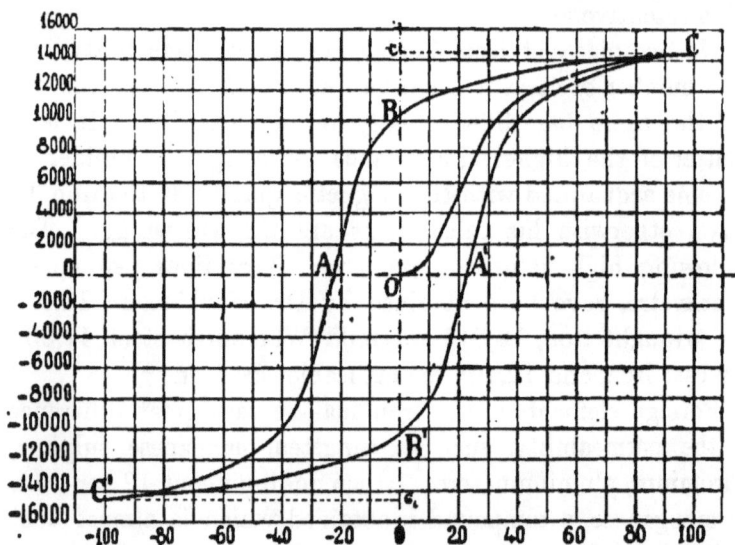

FIG. 17.

the iron has no magnetism. Upon the magnetising force, H, increasing, as indicated by horizontal values, the magnetic induction, B, also increases, as indicated by vertical values. This increase stops at point C, thus giving the curve O C for the rise of magnetism in the iron. The magnetising force, H, which is now 100, is gradually decreased, and so the iron becomes demagnetised ; when H is brought back to zero it will be found that, instead of B being also brought back to zero, it has a value of 10,000, as shown by the point B on the demagnetising curve, C B. As the current now changes its direction so the magnetising force, H, also changes its direction, and this is shown by negative values plotted to the left of the zero line. This value of B—namely, 10,000— is what is termed the "residual magnetism" of the iron, and its amount varies considerably according to the nature of the iron. Hard iron, as steel, holds on to its residual magnetism, while soft iron, as wrought iron, easily parts with it. To get rid of it, the iron must be magnetised again by sending the current the opposite way round the iron, thereby giving the magnetising force a negative value, and this is precisely what an alternate current does ; so that the first effects of this negative magnetising force are to neutralise or kill this residual magnetism, which is shown by that part of the curve from B to A. At a value of − 24 for H, the iron is entirely demagnetised, for the value of B is then brought down to zero. The iron is now again magnetised by the gradually increasing negative values of H until the point C' is reached, which is equal to C ; upon decreasing the values of H, and so again demagnetising the iron, there is again residual magnetism left, because, when H is at zero, the value of B is 10,000, as shown by B' in the curve C' B'. The current, changing its direction a second time, brings values of H positive again, and so the rest of the curve B' A' is formed by the first values of H. The above forms a complete magnetic cycle of four operations, each cycle consisting of two magnetising and two demagnetising operations, as indicated by the closed curve, C B A C' B' A'. On account of this residual magnetism, the magnetic induction, B, always lags behind the magnetising force, H, as can be seen in Fig. 17, for when H = 0, B = 10,000 ; the same when the current direction is changed. The whole of the area enclosed by the closed curve, C B A C' B' A', measures the waste of energy that takes place due to this hysteresis ; this wasted energy

L

appears in the form of heat, consequently a bar of iron that is magnetised by an alternating current so that its magnetism is being constantly reversed, will rapidly heat up to a considerable degree. The greater the number of reversals per second, the greater will be the loss of energy in proportion, and the greater the magnetic induction the greater the loss also, but not in proportion—100 per second may be taken as a common practical number of cycles for an alternating current, and putting the magnetic induction, B, at 15,000, we may say that for soft wrought iron there is a loss of 1·5 watts for every cubic inch of metal, or about 16 h.p. for every ton of iron used. These figures will give some idea of the great amount of energy that is dissipated in heat in masses of iron when acted on by alternating currents. Although there are several other detrimental effects besides residual magnetism which are together classed under the name of hysteresis, yet residual magnetism is the most important factor. In very soft annealed iron this residual magnetism may reach as much as 90 per cent. of the maximum induction—that is, upon withdrawing the magnetising force, 90 per cent. remains behind, so that the iron is almost as strong a magnet as when the current was flowing round the magnetising coils. But this great amount will quickly disappear upon the slightest vibration of the iron, such as giving the metal a smart tap, or heating it ; both the tapping and the heating cause the molecules to be put into an increased activity, and this destroys the magnetism. Application of great heat causes a permanent magnet to entirely lose its magnetism, and this stage is reached when the iron becomes red-hot, and upon cooling down the metal does not regain its magnetism.

The Magnetic Circuit.

In expounding the electrical circuit we had recourse to a water analogy. Here, in expounding the magnetic circuit, we shall have recourse to the analogy of the electric circuit. That which tends to cause a flow of magnetic lines of force is named the magneto-motive force, even as E.M.F. tends to move electricity ; this flow of lines is named the magnetic flux, and so corresponds to the current in the electric circuit. The obstruction that is offered to this flow of lines is named the magnetic resistance, and so has its analogue in resistance of the electric circuit. The fundamental law of the magnetic circuit is similar to that of the electric circuit, because the magnetic resistance is a ratio between magneto-motive force and magnetic flux. We have thus the following analogous laws :

For the Electric Circuit.

$$\text{Current} = \frac{\text{Electromotive force}}{\text{Resistance}}.$$

For the Magnetic Circuit.

$$\text{Magnetic flux} = \frac{\text{Magneto-motive force}}{\text{Magnetic resistance}}.$$

Magnetic induction, denoted by the letter B, signifies the number of lines per square centimetre, and the total induction, or the magnetic flux, denoted by the letter N, signifies the total number of lines, or the value of B multiplied by the cross-sectional area in square centimetres.

The magnetising force per centimetre, or, as it is technically called, the magneto-motive force, has the same value as the letter H, while we may write R for magnetic resistance. We have thus the following formula, expressive of the first law of the magnetic circuit :

$$N = \frac{H\,l}{R}.$$

where l signifies the length of the circuit in centimetres.

L 2

Since iron is the only substance that conducts magnetic lines well, all magnetic circuits may be said to consist of air, iron, or partly air and partly iron. A solenoid is an instance of the first, a ring of iron of the second, and a bar or horseshoe electromagnet of the third.

Magnetic lines of force tend to take a somewhat circular course, so that if the magnetising coils be wound so as to make a circular helix or circular solenoid, most of the lines will flow in the circular hollow of the helix, scarcely any leaking out or taking stray paths. If this hollow be filled with iron, thus giving a circular electromagnet, the lines of force, and consequently the magnetism, will be confined to the interior mass of the iron. The result will be rather peculiar, for owing to this, the iron ring is useless as a magnet—no magnetism being evident. But cut through the ring, breaking the metallic circuit, here we get an air-gap across which the lines of force flow; therefore, in this air-space between the two ends of the ring, a magnetic field exists, and its magnetism is available. On account of this air-gap, lines will leak out or curve out of the direct path which they would have taken had no gap existed, and the wider this gap is, the greater is this leakage. Also, the more the circuit departs from the circular form, the greater is the leakage. In a circuit—say, of a rectangular form—where right-angled corners exist, the leakage is very high, the lines leaking out into the air in dense bunches near the corners. All corners and sharp angles should, therefore, be avoided in magnetic circuits, for their presence is as detrimental as bad insulation is in an electrical circuit. In dealing with magnetic questions, the term "poles" is freely used, and remarks made respecting them are mostly confined to merely theoretical technicalities, which are seldom wanted in practical work. The poles of a magnet simply mean that part where the magnetic lines of force leave and that part.

where they enter the iron, the former being named the "north pole," and the latter the "south pole." The direction of flow of magnetic lines is determined by the direction of flow round the iron of the current which magnetises the iron. The direction of flow of current is arbitrarily fixed, as before mentioned, consequently the direction of flow of magnetic lines must likewise be arbitrarily fixed; and whatever the direction of either may be, reversing the direction of current reverses the direction of the lines. Knowing the direction of current in the wire (from positive to negative terminal), the polarity of a magnet is easily found; wind the wire round the iron in a direction going from right to left, or with the sun, or clockwise, beginning at the end nearest yourself and finishing at the far end, then, if the current enters at the near end, and leaves at the far end, so as to circulate round the iron in the way a right-hand corkscrew would work into a cork, this far end where the current leaves the helix, or the end towards which the corkscrew would travel, would be the north pole, and the near end where the current enters, would be the south pole, because the direction of flow of magnetic lines of force would be along the iron from the near end to the far end. In the case of a complete iron circuit, like a ring of iron, there are no poles, because no lines of force emanate from the iron; if the circuit be broken, the two ends become poles. A straight bar electromagnet has its poles at its two ends; if the bar be divided into two, at the place of division two more poles will appear, and as the lines leave one end to enter the other end in flowing across the air-gap, therefore, at any break or discontinuity of the iron a north and a south pole would appear, and however many times the magnet may be divided up, each piece will yield two poles.

The permeability of air remains constant at unity, irrespective of the intensity of the magnetic induction, which is

very different from the great fluctuations of the permeability
of iron, and owing to the very high magnetic resistance
which air offers to the propagation of magnetism, the air-gap
in a magnetic circuit is kept as small as possible; this air-
gap contains the magnetic field, and in the magnetic circuit
of a dynamo it is made as strong as possible, consistent with
economical features, and its length limited, so that there is
only just room enough to permit of the armature wires being
rotated

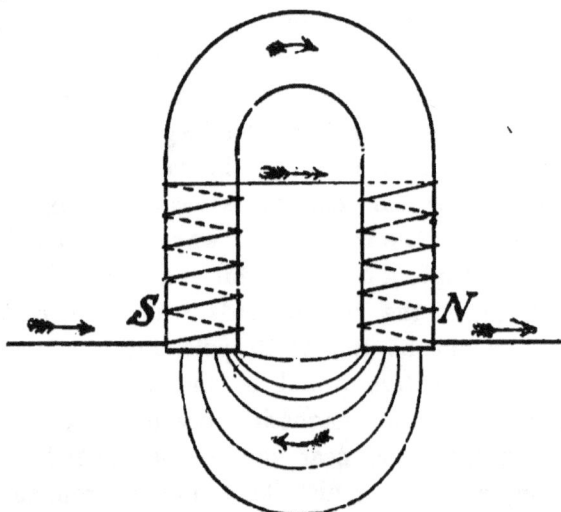

FIG. 18.

If a straight bar electromagnet be considered, its air
return path from north pole to south pole is enormous in
length, being much longer than the straight iron portion of
the circuit. By bending the bar into a horseshoe shape, as
shown in Fig. 18, its air path is considerably reduced; this
is named a "horseshoe electromagnet," and is the outline of
the general shape given to the magnetic circuit of a dynamo.
It will be observed that the path taken by the lines across
the short air-gap from one pole to the other is not straight,

but curved, only a few lines flowing at right angles to the bar. The magnetising coil is shown circulating round the bar in a right-handed direction, beginning at the left leg, at which place the current is assumed to enter the coils; consequently, the far end, or right leg, is the north pole. The direction of flow of the magnetic lines is shown by the two arrows. A hint may here be given that when winding the two legs of a horseshoe magnet the direction of winding is *reversed* when crossing over from one leg to the other, on account of the bend in the iron; in imagination, carry the winding of Fig. 18 round the bend, when the reason for this will be seen.

To concentrate the magnetic field between the two poles, masses of iron, called "pole-pieces," are attached, having their opposing faces curved so as to almost embrace a circle; this tends to direct the flow of lines more in a straight line from one pole-piece to the other. In this form the working of the iron would be very troublesome and therefore expensive, and, in addition, the coils could not very well be wound on the legs of the magnet. To obviate these drawbacks the magnet is usually made in three pieces, the bended part being replaced by a straight bar, called the "yoke," and screwed on to the two separate legs, or "limbs," each limb and pole-piece being made from one piece; it can now be understood where the advantage comes in by using cast iron in place of wrought iron, for the limb and pole-piece can be cast at once into the design required, and only require a little planing and machining up. The magnetising coils are carried on the limbs of the magnet; it is of the utmost importance that the yoke should be screwed on the limbs in such a way that a good and true metallic connection is made, because if the surfaces are not planed with great accuracy so as to get as many contact points as possible, the thin film of air between the surfaces will add greatly to the magnetic resistance of the circuit. This means that joints should be avoided, and limited to what is absolutely necessary for the economical turn out of the magnet. When the continuity of the iron is broken by a joint, the latter forms an obstruction to the magnetic flow; a smooth-surfaced joint, tightly made, is equivalent to an air-gap of ·0013cm., or a length of iron of ·4cm. When wrought iron is used

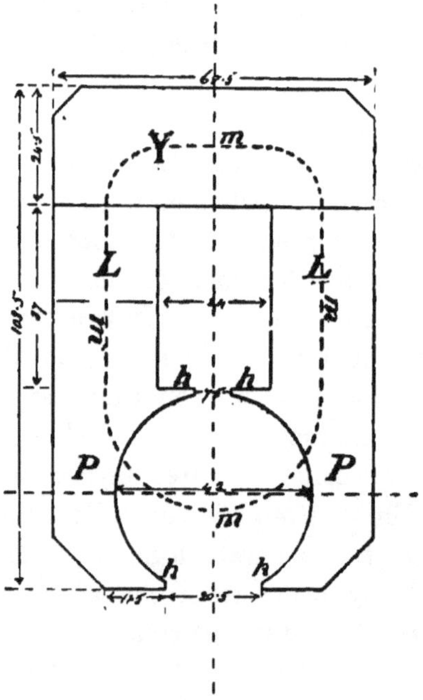

FIG. 19.

for the magnets, the pole-pieces are often made of cast iron, and then screwed on to the limbs of the magnet. This slightly lowers the magnetic flow, but the advantages of having pole-pieces cast into shape compensates. All corners should be curved down or cut off, otherwise great magnetic leakage will occur. It is very usual to have the yoke like-wise made of cast iron. When this is so, the yoke is made larger in sectional area than what it would be if made of wrought iron. The magnetising coils can be wound by putting the magnet limbs in a lathe. When both are wound, the two coils are joined together. The winding takes place on a brass bobbin fixed on the magnet limb. Often, how-ever, the wire is wound on the brass bobbin alone, and the latter slipped on to the limb, which is then screwed into the yoke. The modified arrangement is shown in Fig. 19, which may be taken as a good example of the field magnets of a dynamo, the type being after that used by Messrs. Siemens Bros. A dynamo made with this size field magnets would give an output of about 45,000 watts at a speed of 450 revolutions per minute, and be capable of maintaining 800 incandescent lamps of 16 c.p. each. The drawing is on a scale of $\frac{1}{20}$, the measurements given being in centimetres, the depth of the magnets being 48cm. The thick dotted line denoted by m, m, m, m, traces the magnetic circuit; Y denotes the yoke ; L L denote the magnet limbs ; P P denote the pole-pieces; h, h, h, h, denote the horns of the pole-pieces. The depth of winding on the limbs would be about 9·5cm.

In working out problems on the magnetic circuit, the first thing to know how to calculate is the magneto-motive force of a coil. A unit line of force is of such intensity that it would act with the force of one dyne upon a unit pole positioned in its path. Unit intensity of magnetic field exists when there is one line of force per square centimetre. Magneto-motive force per centimetre, or magnetising force, is the same as H, and when the medium through which the magnetic lines flow is of air, the intensity of the field is measured by H also, because the permeability of air is 1—that is to say, if the magneto-motive force per

centimetre be 15, then 15 lines of force per square centimetre will flow when air is the medium. H is dependent
on two things : (1) the strength of the current ; (2) the
number of turns of wire. The product of the above is
named the ampere-turns.

When merely the intensity of H is required, the value of
the total magneto-motive force must be divided by the length
of the circuit in centimetres ; but for calculating the magnetic
flux in a circuit, the total magneto-motive force is required,
and so length is left out. The product ampere-turns can be
made up of any two factors so long as the product remains
constant ; for example, suppose the current were nine amperes,
and that it circulated round the iron 100 times, then the
ampere-turns would be $9 \times 100 = 900$. If the current were one
ampere, and it circulated 900 times round, the ampere-turns
would still be 900, and the magneto-motive force would be
the same as before.

It was stated that unity intensity of field existed when
there was one line of force per square centimetre. A unit
magnetic pole signifies a centre of magnetism, such that all
round it on every side there is a unit intensity of field at
a distance of 1cm. from the centre. This enclosing space
is evidently of a spherical form, the pole being in its
centre, and as the radius or distance between the centre
and the boundary surface is 1cm., the surface of the
sphere will have 4π square centimetres ; and since there
is unity intensity of field at every point on the surface,
therefore there will be one line of force per square centimetre, and as there are 4π square centimetres, consequently
there must be $4\cdot\pi$ lines of force issuing from the pole, or
magnetic centre.

The above theoretical digression is necessary in order to
explain the complete formula for magneto-motive force.
Other things being measured in absolute or C.G.S. units,

the current must be expressed in absolute units, and that being so, the magnetising effect of one absolute unit of current circulating once round will produce 4 π lines of force. Now, an ampere is one-tenth of the absolute unit of current, therefore one ampere-turn = 4 π ÷ 10 = 1·256 lines, so that

Magneto-motive force = ampere turns × 1·256.

or, $\qquad$ H $l = 1\cdot256$ S i,

where H = magneto-motive force per centimetre ;
$\qquad$ S = number of turns of wire ;
$\qquad$ i = current in amperes.

For example, suppose a bar of iron to be wound with 1,000 turns of wire through which a current of 20 amperes was sent, then the total magneto-motive force would be

$$H\ l = 1\cdot256 \times 1,000 \times 20 = 25,120.$$

The magnetic resistance of the circuit depends on its dimensions and on the value of μ. The former is like the electric circuit, proportional to length and inversely proportional to area. The value of μ depends on the value of B, or density of lines per square centimetre that are likely to flow through the iron ; so that when B is found, μ is found by referring to Tabulation 21 or 22. We have thus the following formula :

$$R = \frac{l}{\mu\ a},$$

where R = total resistance ;
$\qquad$ l = length in centimetres ;
$\qquad$ a = cross-sectional area in square centimetres ;
$\qquad$ μ = permeability.

Where the circuit is composed partly of air and partly of iron, the several portions must be added together for the total resistance.

The magnetic circuit of a dynamo may be divided up into three portions : (1) the field magnets ; (2) the two air-gaps ; (3) the armature core. Here the law would be

$$\text{Magnetic flux} = \frac{\text{Magneto-motive force}}{\text{The three magnetic resistances}},$$

or,

$$N = \frac{1 \cdot 256 \, S \, i}{\dfrac{l_1}{\mu_1 \, a_1} + \dfrac{2 \, l_2}{\mu_2 \, a_2} + \dfrac{l_3}{\mu_3 \, a_3}}.$$

In the electric circuit the current is the same at any point, so in the magnetic circuit the magnetic flux is the same at any point. In the former, the difference of potential between two points is obtained by multiplying the current by the resistance between those two points ; in the latter, the difference of magnetic potential between two points is obtained by multiplying the magnetic flux by the magnetic resistance between those two points.

Since the permeability of air is unity, therefore when the magnetising or magneto-motive force has a value of one C.G.S. unit, there will be a magnetic induction, B, of one line of force per square centimetre, so that for air H signifies both the magneto-motive force and the magnetic induction, B ; or H = B.

The following example will show how to calculate the number of ampere-turns required to give a certain magnetic flux in a circuit of given dimensions. A dynamo requires a magnetic field having a total flux of 2,000,000 lines of force ; how many ampere-turns must be wound on the magnet limbs ? The dimensions of the circuit being as follows, leakage being neglected :

		Length.	Area.
Field magnets	...	100cm.	400 sq. cm.
Air-gap ...	...	1·5 ,,	800 ,,
Armature core	...	20 ,,	200 ,,

The induction, B, per square centimetre through the field magnets comes to 5,000 lines, and signifies a permeability of 2,500. The induction, B, per square centimetre through the armature core comes to 10,000 lines, and its permeability is hence 2,000. The induction, B, per square centimetre through the air-gap comes to 2,500 lines. The value of μ for air being 1.

$$\text{Ampere-turns} = \frac{\text{magnetic flux}}{1\cdot256} \begin{pmatrix} \text{The sum of the three} \\ \text{magnetic resistances} \end{pmatrix}$$

or, $$S\,i = N \left(\frac{l_1}{\mu_1\,a_1} + \frac{l_2}{\mu_2\,a_2} + \frac{l_3}{\mu_3\,a_3} \right) \div 1\cdot256,$$

the first resistance term being that of the field magnets, the second that of the two air-gaps, and the third that of the armature core. Inserting values, we have

$$S\,i = 2,000,000 \left(\frac{100}{2,500 \times 400} + \frac{2 \times 1\cdot5}{1 \times 800} + \frac{20}{2,000 \times 200} \right) \div 1\cdot256$$

$$= 2,000,000\ (\cdot0001 + \cdot00375 + \cdot00005) \div 1\cdot256\ ;$$
$$= 2,000,000 \times \cdot0031 = 6,200.$$

So that 6,200 ampere-turns would be required.

CHAPTER IV.

Generation of Current—Field-Magnet Winding—Armature Winding—Working in Parallel—Notes on Running—Cost and Output of Dynamos—Electromotors.

Generation of Current.

All dynamo-electric machines consist, broadly, of two parts. The first part may be called the "magnetic circuit," and the second part may be called the "electrical circuit." The first comprises the two magnet limbs, the yoke, and the pole-pieces, called collectively the field magnets; to this must be added the iron core of the armature. The second comprises the armature wires, the brushes, and the conducting wires, lamps, etc. We have thus to deal with two separate and distinct circuits—one being magnetic, the other being electric. To produce a current of electricity in the closed electrical circuit, part of this circuit must be put into motion, so as to cut through part of the magnetic circuit ; this moving part of the electric circuit is named the "armature," and the part of the magnetic circuit which it cuts through is evidently the magnetic field existing in the air-space between the two poles of the field magnets. It does not signify which circuit moves, or which is stationary, but one of the two must move. Some dynamos are built in which the field magnets move round whilst the armature remains stationary, but it is more the custom to rotate the armature, and this custom will be described here.

Since the armature part of the circuit rotates, sliding connections, called "brushes," join on to the rest of the circuit, consisting of the conducting mains, switches, lamps, etc. The iron core of the armature serves two purposes— it firstly, and chiefly, assists to fill up the air-gap inside of the armature ring or drum upon which the wires are wound, and so provides a path for the magnetic lines; secondly, it acts as a mechanical structure necessary for the proper strength of the machine; so that, although rotating with the armature, it forms part of the magnetic circuit.

An armature is built up with a great number of wires, which are wound round an iron core keyed on to the shaft, which is driven either by belting or else direct by being coupled up to the steam-engine shaft. A brief explanation as to the way various kinds of electric currents are generated in the armature will now be given.

Electric currents produced by dynamo-electric machinery may be classed under four distinct heads as follows : (a) alternating currents ; (b) polyphase currents ; (c) pulsating currents ; (d) continuous currents.

Every dynamo, by virtue of the rotation of a conductor in a magnetic field, gives, in the first place, an alternating current, and the other kinds are simply modifications of it, as will be noticed later on. The currents used at present for distributing light and power are either alternating or continuous. With regard to polyphase currents, this system being new, is mostly confined to experimental plants. Concerning pulsating currents, this kind is at present very limited in its application, but within the last year or two has been found useful in working reciprocating electric machines, such as percussion rock drills, etc.

(a) *Alternating Currents.*—The armature will be limited to a single coil in order to explain the action more easily. Imagine this coil to occupy a vertical position, the lines of force threading through it at right angles. Let the coil be moved counter clockwise through an angle of 90deg., or a

quarter of a revolution.   Just at the point of starting, the
coil will move almost parallel along the lines, and so the rate
of cutting will be at the lowest, whilst when the coil is at
right angles to its first position the rate of cutting will be
at the greatest, because the coil is moving at right angles to
the lines, and so will cut right down and through them.
During the recond quarter revolution, the rate of cutting
will decrease, until on completing one half a revolution, or
180deg., the coil will again slide along the lines, and so will
do no cutting, as happened at the starting.   The other half
revolution the same thing is repeated, but with this important
change : the first half revolution the coil travelled in a
downward direction, and in the second half it will travel in
an upward direction.

The number of lines of force cut per second determine the
value of the E.M.F. set up in the armature coil, and the
number cut in one complete revolution will vary from a
minimum to a maximum, then to a minimum, then to a
maximum, and then to a minimum again ; therefore it
follows that the E.M.F. set up will rise and fall in a similar
way.   The direction in which the E.M.F. acts depends on
whether the lines of force are cut by a wire having a downward motion, or by one having an upward motion.   This
being so, the E.M.F. acts in one direction during the first
half revolution, and in the opposite direction during the
second half ; as the current produced depends on the E.M.F.,
therefore we have a rise and fall of current with the current
flowing in one direction, and then another rise and fall of
current with the current flowing in the opposite direction.

The first quarter revolution, from vertical position of
0deg. to 90deg., the current rises from zero to a maximum
value ; during the second quarter, from 90deg. to 180deg.,
the current falls to zero ; during the third quarter, from
180deg. to 270deg., the current again rises from zero to a

maximum ; and during the last, or fourth, quarter from 270deg. to 360deg., it again falls to zero. The above forms a cycle, which is graphically illustrated by the top curve in Fig. 20, where the horizontal values refer to the angles of rotation, or position of the coil, and the vertical values give the E.M.F. at those positions. That part of the curve above the horizontal axis denotes one direction of the current, say, that arising from the first half revolution, say, the positive direction, while the part below denotes the other or negative direction, produced by the second half revolution. The two ends of the armature coil are attached to two brass rings, one to each, insulated from one another, and fixed on the armature shaft. On each ring a brush rests, and as they revolve with the armature the brushes make a sliding contact and so conduct the current into the external circuit, to the lamps or wherever else it is required, the current leaving by one brush and then returning by the other.

It is usual to express the horizontal values of the above curves in fractions of seconds of time : suppose one complete revolution is made in the time ·1 of a second, then the distance T_0 T_1 represents ·1 of a second, because the whole curve is the result of one revolution. The E.M.F., therefore, at the commencement, or at time 0, is 0, at ·025 of a second it is at a maximum, so that it takes ·025 of a second to grow from nothing to its full strength. The distance T_0 T_1 measures what is technically named the "periodic time," or the length of time that is occupied to enable the E.M.F. to rise and fall, first in a positive and then in negative direction. This cycle of fluctuations limited by T_0 T_1 is termed an "alternation"; hence, in a machine having only two magnetic poles, there can only be one alternation per revolution, which is the case shown by Fig. 20, and assuming that the armature revolves at 600 revolutions per minute, this will give us 10 alternations per second, because it makes

10 revolutions per second. If the machine had four poles, then the current would have two alternations per revolution, and hence with the above speed 20 alternations per second. And however many poles the machine may have, the number

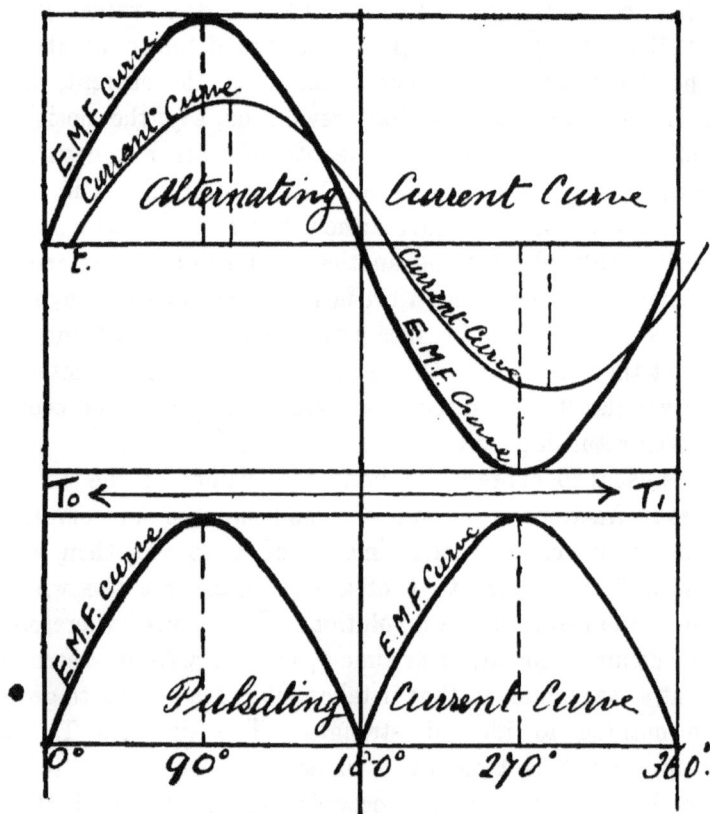

FIG. 20.

of alternations per second, or its "periodicity," is obtained by multiplying the number of revolutions per second by half the number of poles.

It must be well understood that an alternation does not

mean a mere reversal in direction, say, when passing from a north pole to a south pole ; what it does signify is a complete cycle, consisting of a positive wave and a negative wave, so that the top curve in Fig. 20 represents one alternation, as explained above. The symbol used to express this cycle is ⌒⌣, being a diminutive curve. It may be observed that the E.M.F. curve of Fig. 20 represents a curve of sines. Now it has been found by careful experiments that the values of the E.M.F. of a revolving coil rise and fall very nearly in the same way as the sine of the angle which the coil has in its various positions. For example, when the coil is perfectly vertical it is at 0deg. or 360deg., and it is moving almost parallel with the magnetic lines, consequently the E.M.F. is at zero, and the sine of 0deg. or 360deg. is zero also ; when at an angle of 45deg., or half-way between vertical and horizontal positions, its value will be found to be proportional to the sine of 45deg., which is $\sqrt{\frac{1}{2}}$ or ·7 ; coming to 90deg., or the horizontal position, its value will be proportional to the sine of 90deg., which is 1, and all intermediate values are also nearly in proportion. To simplify calculations it is customary to consider the curve as a curve of sines, and for most practical work this answers well enough.

The rise and fall of the current depends on the rise and fall of the E.M.F., but alternating currents are somewhat analogous to the increase and decrease of magnetic induction when a magneto-motive force makes a magnetic cycle. It will be remembered' that the induction, B, lagged behind the magneto-motive force, H (see Fig. 17), and putting E.M.F. for H, and current for B, we have a similar lagging of the current behind the E.M.F. in alternating currents, so we may draw a second curve for the current which commences a short time after the curve of E.M.F. ; the current wave thus lags behind the E.M.F wave. This lagging is named a "retardation in phase," and when two waves, such as the two waves of E.M.F., of two machines, coincide with each other, they are said to be "in phase," and when one wave lags behind the other it is "out of phase."

For continuous currents, Ohm's law of the form $C = \dfrac{E}{R}$ holds good, but in alternating currents, although Ohm's law holds good just the same, yet it must be written in a more complex form, because the E.M.F. is constantly changing in value and direction, same with the current; this gives rise to certain important effects, which will be discussed when treating of Ohm's law for alternating currents. For the present, it suffices to say that these effects cause a decrease of the current, hence the current waves are drawn smaller in amplitude than the E.M.F. waves; these opposing effects act as if the copper conductors of the circuit had more resistance than what they really had, and so cause a smaller current to flow. The distance T_0 t marks the "retarda- tion in phase" between the E.M.F. wave and the current wave.

(b) Polyphase Currents.—These currents are produced by generating two or more alternating waves of E.M.F. by having a machine with two or more magnetic fields, which come into action successively. When the generating machine has two pairs of magnets, wound in such a way as to give two distinct magnetic fields, these fields are placed at right angles to one another, and so the two E.M.F. waves follow each other—one wave being 90deg. behind the other—then when the first wave is at a maximum, the second wave is at zero. Two waves like this are said to be "in quadrature," and currents so obtained are named two-phase or di-phase. Similarly, by having three waves of E.M.F., each following the other at an angular or phase distance of 120deg., three

currents are produced, each lagging behind the other 120deg., and these currents are named three-phase or tri-phase.

(c) *Pulsating Currents.*—A pulsating current signifies one which flows always in the same direction, but which rises and falls in value, increasing from zero to a maximum and then dropping to zero again. This action is easily obtained by what is called "commutating" the alternating current. To effect this, instead of joining the two ends of the armature coil to two separate collecting rings, they are connected to a split tube or ring, so that one end of the armature coil is fixed to one semicircle of the tube and the other end of the armature coil to the other semicircle. The two halves being insulated from one another ; the brushes are placed diametrically opposite each other on this split tube, in such a position that when a reversal of direction of current is about to take place the brushes change halves by breaking contact with one half of the tube and making contact with the other half, so that at the moment of reversal of flow the brushes likewise make a reversal of contact, by passing from one portion of the tube to the other. In this way the brushes lead off a current which is commutated in one direction. The split tube is called a " commutator."

The bottom curve of Fig. 20 shows, graphically, the nature of a pulsating current, both waves of the curve being now above the horizontal axis, thus indicating the same direction.

(d) *Continuous Currents.*—If incandescent or arc lamps were run with a pulsating current derived from one coil, the light would fluctuate up and down, on account of the rising and falling of the current ; but by using a number of coils in the armature arranged so that one set comes into action just as another goes out, the curves obtained would overlap one another, and so the crests of the wave-curves would be very close together, and form a wavy line whose resultant crests and hollows would be so small that it would practically amount to a straight line. This effect is shown in Fig. 21. This can easily be traced, for suppose there were two coils, one being fixed at right angles to the other, then when No. 1

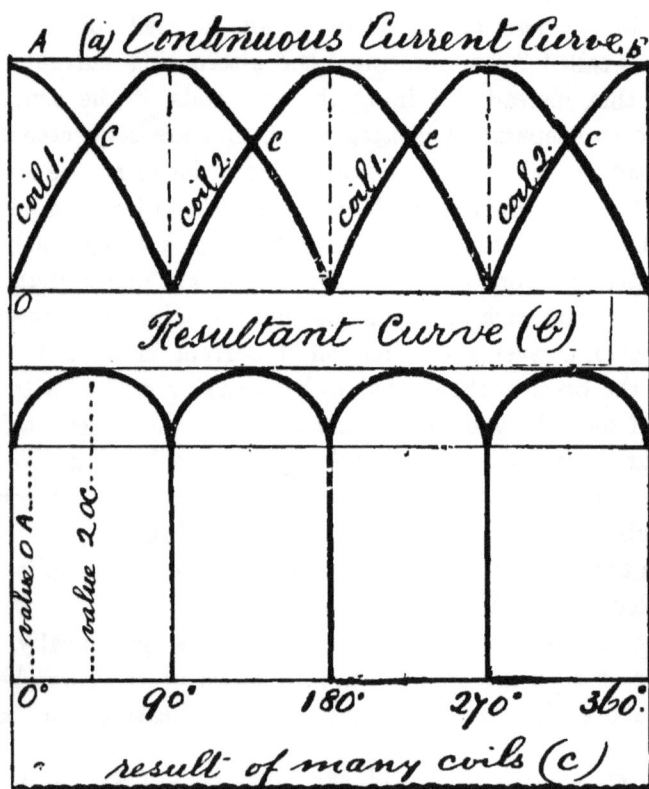

FIG. 21.

would be vertical, or in the position of least activity, No. 2 would be horizontal, or in the position of greatest activity, as shown by the two top curves of Fig. 21 ; and the two waves cross each other at an angle of 45deg., which signifies that when No. 1 coil is half-way between vertical and horizontal

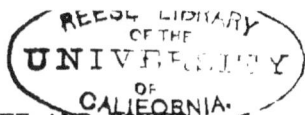

positions, and decreasing in value, No. 2 coil is similarly situated, but increasing in value. This is the relative position of the two coils towards each other. Now let us examine what the resultant E.M.F. wave as obtained at the brush points will be. It is evident that the maximum E.M.F. of either coil alone at the two brushes is that given by the crest of either wave, so we can draw a line, A B, along the top of all the crests, and this line indicates the minimum resultant E.M.F. at the brushes. Now the lowest point reached is fixed by the crossing of the waves, but at these crossing points, C, the E.M.F. of the two coils are equal, consequently the total E.M.F. due to both coils must be double the height of these points, O C, and this is the resultant maximum at the brushes. Intermediate values can be obtained by adding together the vertical values of the two curves at every instant. These integrated values are then plotted again and give the resultant curve, b. Hence the resultant E.M.F. at the brushes fluctuates between the two values, O A and 2 O C. By increasing the number of coils to four, we get a resultant curve whose fluctuations are more limited, and using more coils we can reduce the fluctuation so that it resembles a wavy line, as shown by c.

The commutator which was used for leading off the pulsating current must now be supplanted with one divided up into a number of parts. In addition to this multiplicity of coils, all the coils are joined up in series, so that the end of one is joined to the beginning of the next and so on. The beginning of each coil, and therefore the end of the next coil, are connected to one part or "segment" of the commutator, as is clearly shown in Fig. 22, similar numbers referring to each coil and its proper segment. In this figure, s signifies the driving shaft; c the commutator, divided into segments by insulation denoted by thick lines; a the armature core, upon which are wound the eight coils; S and N signifying

the south and north pole-pieces of ·the field magnets, the
rotation of the armature being counter clockwise, or against
the sun.   The leading-off wires from the two brushes, B and
B¹, are not shown in order to make the diagram clearer.

Now, the E.M.F. set up in each individual coil depends on
the position which that coil occupies in the magnetic field.
No. 1, for example, lies at the top, parallel to the lines of
force, consequently at this point its rate of cutting will be at

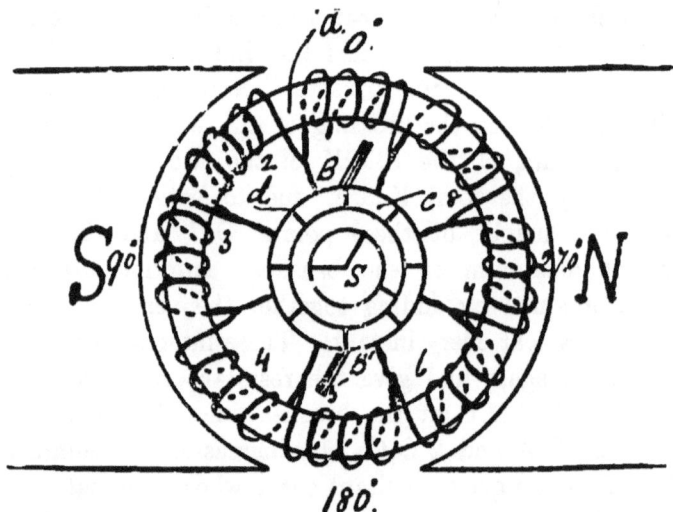

FIG. 22.

zero, or a minimum ; No. 2 is 45deg. further on, and its rate
will be higher ; while No. 3 lies at right angles to the lines
of force, and its E.M.F. will therefore be at a maximum,
because it is making a straight downward motion across the
lines.   The same argument applies to all the rest.   The effect
of all the coils being joined up in series is that the E.M.F.'s
of all lying between 0deg. and 180deg., or on one half of the
armature (Nos. 1, 2, 3, 4), will be added together.

In a similar way the E.M.F.'s of all the coils of the other

half (Nos. 5, 6, 7, 8) will be added together, but the coils belonging to the first half are in parallel with the coils belonging to the second half, because they form two paths between the two brushes, B and B^1. The sum of all the E.M.F.'s of the first half is equal to the sum of all the E.M.F.'s of the second half, and this is the same thing as there being only one E.M.F., and each half has the same resistance and is in parallel ; consequently the resistance between B and B^1 is one half the resistance of the sum of the resistances of either half, or equal to one quarter of the resistance of all the coils counted all round the ring.

Upon examining Fig. 22 carefully, it will be seen that the brushes are made so broad that they touch across two segments of the commutator. This is done in order that they shall make contact with one segment before they leave the other, and so not cause any break in the circuit in changing over ; the result of this is to complete the circuit of the coil that happens to have its ends on these two segments. This is called " short-circuiting " the coil. Now it is evident that if the coil were short-circuited by a brush when that coil was in the position that No. 3 coil occupies in Fig. 22 (or the position of greatest "activity," as it is called) a very heavy current would flow for a moment through the coil, due to the high E.M.F. which is being set up in it and the low resistance of the coil ; this current would probably burn up the insulation of the wire, and cause such sparking at the brush that both it and the commutator would suffer terribly. It is therefore necessary that the change from one segment to another should take place at a point where this is avoided ; such a point is the one where the brushes are shown in Fig. 22. Here the coil is moving mostly parallel with the lines of force, its own E.M.F. is almost zero, hence short-circuiting takes place with impunity, and the brushes can collect the integrated E.M.F.'s of all the other coils, from the coil having

the highest to that having the lowest, since these coils are not
under the brushes, and hence cannot be short-circuited, and
their circuit is only completed through the external or lamp
circuit.

Field-Magnet Winding.

Continuous-current dynamos have their field magnets
wound in three distinct ways, which may be classed as
follows : (a) series wound ; (b) shunt wound ; (c) compound
wound. Each kind has its own particular use.

(a) *Series Wound.*—A machine is "series wound" when
the whole current is caused to circulate through the field-
magnet coils, either before it goes off into the lamp or
external circuit, or afterwards, so that the magnet coils are
simply interposed in the circuit as if they were part of the
main conductor, and this part were coiled round the magnet
limbs. Such an arrangement is shown in diagram (1) of
Fig. 23, where A signifies the armature, B B¹ the positive
and negative brushes, F the field-magnet coils, and l lamps
in series in the external circuit.

From this it is seen that, starting at the brush, B, the
current first passes through the magnet coils, F, along the
main conductor, through the several lamps, l, and back by
brush B¹, thus completing the circuit.

It is a matter not only of great interest, but one of most
vital importance in studying the action of dynamos, to
possess some simple means of a concrete nature to forcibly
illustrate the performance of a machine, by clearly exposing
its action under different conditions of working, such as how
the current and pressure vary ; how speed affects the self-
regulation ; how to find the maximum output or horse-power
for a certain speed, or greatest economy ; how to find the
maximum efficiency, etc. All these valuable things, and
many more, can be obtained by plotting out curves. These
curves, in short, tell us all about the machine, and, as

Prof. S. P. Thompson says, "The characteristic curve stands, indeed, to the dynamo in a relation very similar to that in which the indicator diagram stands to the steam-engine. As

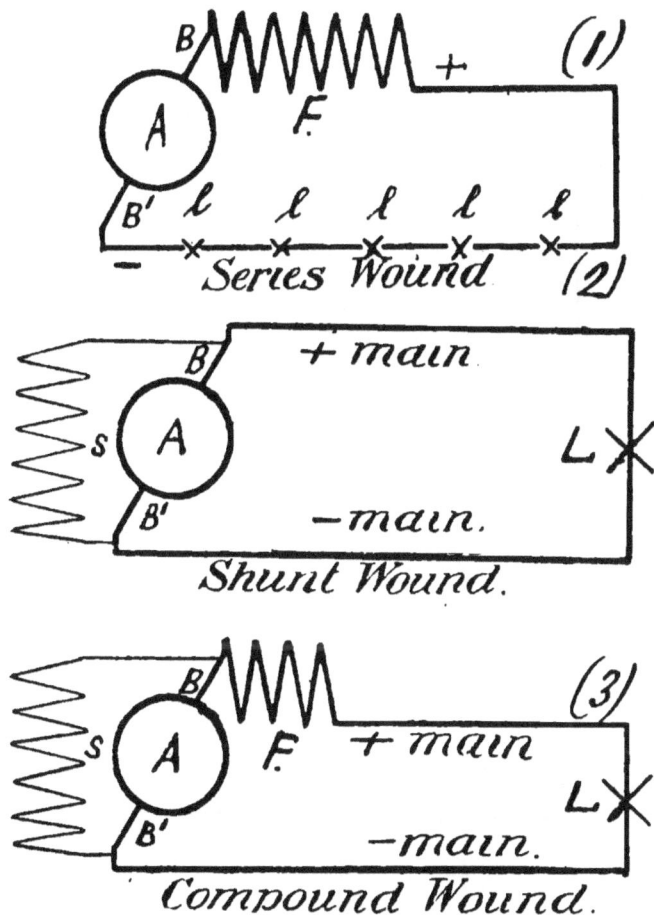

FIG. 23.

the mechanical engineer, by looking at the indicator diagram of a steam-engine, can at once form an idea of the qualities of the engine, so the electrical engineer, by looking at the characteristic of the dynamo can judge of the qualities and performance of the dynamo."

The most important and useful of these curves are those named "characteristic curves." Fig. 24 shows the charac-

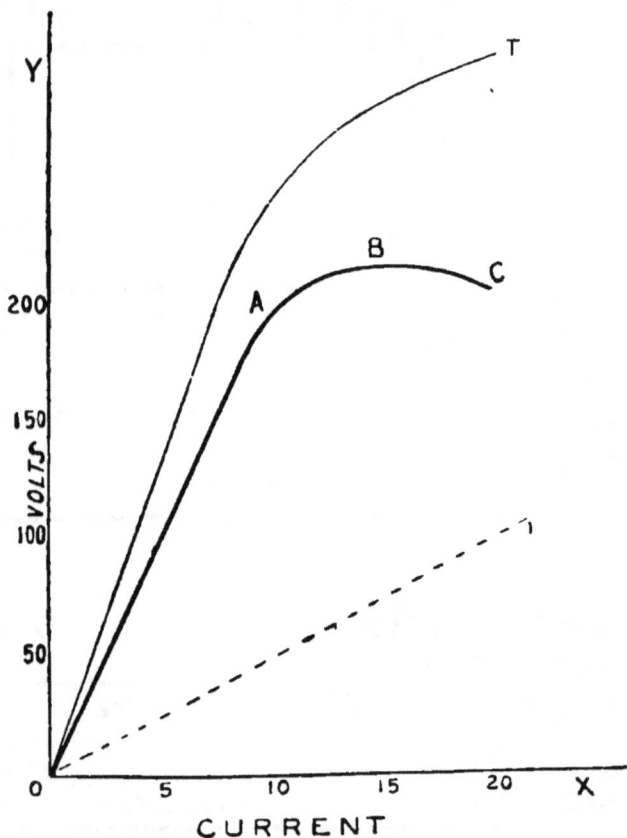

CURRENT

FIG. 24.

teristic curve of a series-wound dynamo, where the E.M.F. is plotted vertically in volts, and the current is plotted horizontally in amperes. As explained, with a series machine there is only one circuit, and therefore only one current which flows round and has the same strength at every point. The way in which the varying values are obtained is by experimental tests conducted in the following manner : The dynamo is first speeded up to a certain suitable speed, the brushes being on the commutator, and the external or lamp circuit is disconnected from the machine : the result is that the external circuit is open, and hence has infinite resistance. . Under these conditions no current can flow, but we obtain a very feeble E.M.F., due to the residual magnetism in the iron ; so our starting point in the curve gives us 0 for current, and almost 0 for E.M.F., as can be seen on referring to the curve O T in Fig. 24. The next step is to close the circuit, but at the same time to insert a very high resistance in the circuit. This gives us a feeble current, which, circulating round the field coils, augments the strength of the magnetic field. The effect of this is to produce an increase of E.M.F. in the armature coils. By decreasing the inserted resistance we allow more current to flow, the field becomes stronger, and consequently the E.M.F. will again rise, and so we can obtain a number of values of current and E.M.F. by gradually decreasing this resistance. We will now examine the curve and see the result of these experiments. The curve rises very rapidly at the commencement, showing that for a small increase of current, say from 0 to 7·5, we have a very great increase in E.M.F., say from 0 to 200. From here the pressure rises more slowly, so that when the current is made three times as strong, say from 7·5 to 20, we only obtain one-and-a-half times the pressure, say from 200 to 300. Beyond the last point any increase in the current gives only a very small increase of pressure. This is because the iron of the field magnets become saturated, and hence the magnetic field increases very slowly. When the difference of potentials at the dynamo terminals is plotted instead of the total pressure or E.M.F. we obtain a curve,

O A B C, below the curve O T, and the difference in heights between the two curves measures the loss of volts due to the resistance of the armature and field-magnet wires—this last curve is called the external characteristic, and is the more useful of the two, as it shows the available power of the machine and how it affects the lamp circuit. The curve O I represents these differences of heights, and hence the drop of volts inside the machine for various currents. The E.M.F. of a dynamo depends on the number of lines of force cut per second. It was stated that the practical unit of E.M.F. of the volt was equal to 10^8 absolute units, and since one line of force is measured by one dyne or absolute unit of force, therefore one volt will be produced when 10^8 lines of force are cut per second. The number cut is dependent on three things : first, the speed of the armature coil ; second, the number of coils in the armature ; third, the magnetic flux through the armature. And for two-pole machines the E.M.F. obtained can be found by the following formula :

$$E = \frac{N\,C\,n}{10^8},$$

where E = number of volts of E.M.F. ;
    N = magnetic flux ;
    C = number of conductors or turns on armature ; and
    n = number of revolutions per second.

For multipolar machines, the number of conductors or turns on the armature must only be counted from one north pole to the next north pole.

The current given by a series dynamo is obtained by dividing the E.M.F. by the total resistance of the circuit. This total resistance comprises several resistances in series. The resistance of the armature wires and the field-magnet coils are together called the "internal resistance"; the resistance of the main conductors, lamps, etc., are together called the "external resistance." Take the case of a series dynamo running a number of arc lamps, the circuit being as in Diagram (1), Fig. 23, where the resistance of the armature A = ·5 ohm, of the field-magnet coils F = ·7, and the resistance of the mains and lamps = 100 ohms. The E.M.F. being 1,000 volts, the total resistance will be ·5 + ·7 + 100 = 101·2, therefore the current will be 1,000 ÷ 101·2 = 9·88 amperes. Series-wound dynamos are only used for running lamps in series, mostly arc lamps, and so they are built to give out a high pressure, say, 1,000 to 2,000 volts, with a small current of 10 to 15 amperes. The lamps being connected up in series, offer very great resistance, hence the high pressure necessary, while since the same current passes through all, only a small one is required. To enable a series dynamo to run lamps in series a regulator is required, otherwise it would be impossible to turn on or off any lamps without interfering with the others. For example, imagine a series machine were put to work with a number of incandescent lamps connected up in parallel, if one lamp is turned out, the resistance will increase, hence the current will decrease, less current will flow through the field coils, the E.M.F. will then decrease, and the current still further. If, on the other hand, a lamp is switched on, the current is increased more than is necessary ; if, running arc lamps in series, a lamp is switched off, the resistance decreases, and the current increases, which would be damaging to the rest of the lamps.

(b) *Shunt Wound.*—The word "shunt," in electrical parlance, signifies usually a branch path of higher resistance than the main path—or a sort of go-by—and has probably been derived from the expression "shunting" used in railway work ; and a shunt current means some fractional part of the main current. A shunt-wound dynamo means that the field-magnet coils form a shunt path to the main circuit, so that when the whole of the current leaves the brush there are two paths open to it, as seen in Diagram (2), Fig. 23, one being the main conductor leading to the lamps, and the other being the coils of the magnets ; the two paths unite again at the negative brush, B^1. Now, the current will divide inversely to the resistance of the two paths, as was explained previously, because these two paths are in parallel; and as only a small part of the current is required to flow round the magnets, therefore the " shunt coils," as they are called, are made of a great number of turns of thin wire, thereby offering great resistance and so only allowing a small current to flow through. The remainder of the current, which forms nearly the whole, flows through the thick conductors to the lamp circuit. So, whereas we have a thick wire to carry the whole current for the series coils of a series machine, we have a thin wire to carry part of the current for the shunt coils of a shunt machine.

A shunt-wound dynamo is probably the best kind that can be used for running incandescent lamps in parallel by continuous currents. There is only one objection to its use, and that is, it is necessary when the lights are subject to being switched on and off, to have someone to regulate the machine by throwing resistance in and out of the shunt coils in order to keep the pressure constant. Suppose a number of lamps working in parallel were switched off, this would increase the external resistance, hence more current would flow through the shunt coils and less through the lamp circuit ; but the effect of sending a greater current round the shunt coils will be to increase the pressure, and when lamps are run at a higher pressure than what they are made for they soon burn out or break. On the other hand, by switching more lamps on more current will flow through the lamp circuit and less through the shunt, and so the pressure will fall, and, as a result, the lamps will burn dull.

Diagram (2), Fig. 23, gives a view of the connections of a

shunt dynamo, the letters denoting the same parts as in Diagram (1).

Shunt dynamos are always used, when possible, for charging secondary batteries, and for this reason : if a series machine were used, there is the danger of having the polarity of its field magnets reversed in the event of the batteries

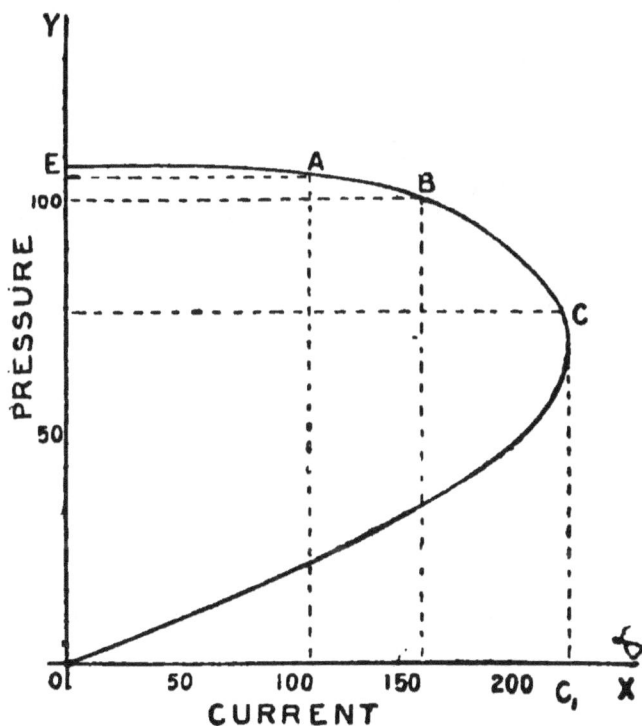

Fig. 25.

sending back their current into the dynamo. With a shunt-wound machine the polarity cannot be reversed because, whether flowing from the cells or from the armature, it reaches the positive brush, and so passes through the shunt coils in the same direction.

Fig. 25 shows the characteristic curve of a shunt-wound

dynamo, which is very different to that of a series machine, the conditions of the two machines on open circuit and short-circuit being exactly opposite. With a series machine, when the main or external circuit is open, there is no current flowing, neither is there any pressure, but with a shunt machine, when the external circuit is open, the only path left for the armature current is through the shunt coils, hence the pressure is high. Taking the other extreme, when a series dynamo is short-circuited by connecting a copper main across its terminals, then an enormous current will flow, and the pressure will also be high ; but short-circuit a shunt dynamo in the same way and it will be found that the pressure drops down almost to nothing, because all the current goes into the external circuit, and none scarcely through the shunt coils, consequently the current then falls to nothing likewise. Having the external circuit open we have no current flowing in it, but a high E.M.F. at the terminals, shown by O E, and this condition gives us the first point of the curve, which is at the top. As the resistance of the external circuit is gradually decreased when closed, so more current flows through it, and consequently less current through the shunt coils, and the result of this is that the magnetic field becomes weaker and the E.M.F. falls in value. This is shown by the falling of the curve as the current increases. It will be observed that the rate of falling is very far from being constant : at the beginning it is extremely slow, but further on past B it is very rapid, until a very little increase of current in the external circuit produces an enormous fall in pressure, the fall being perpendicular, as shown at C. It may readily be imagined that this is due to the way the iron loses its magnetism, for when the magnetic field is weak the permeability of the iron is very high, and a small decrease of the magnetising current will make a great decrease in the strength of the magnetic field, and hence in the E.M.F.

With the series dynamo the opposite holds there : the iron gets saturated, due to the great current that flows through its magnetising coils; and the permeability being very low, any increase of current in the external circuit, and hence in the magnetising coils, will not make the magnetic field much stronger, consequently the E.M.F. rises very slowly, and its rate of rising becomes less and less. The curves of magnetisation of iron should be studied together with the characteristic curves of series and shunt dynamos, since each will help the other. Continuing the explanation of Fig. 25 : at point C the current reaches its maximum value, and if the external resistance is decreased any further, the current will begin to *decrease*, because the field has become too weak to produce a strong current even with a very low resistance in the external circuit ; at the same time, the pressure continues to fall gradually, until, when the machine is short-circuited, both pressure and current disappear, so that the curve takes both a downward (for E.M.F.) and a backward (for current) course. Both the current and the E.M.F. fall at equal rates, as shown by the straight part of the curve, which travels towards zero. Much more can be learnt by carefully studying these curves step by step than by any more written explanations ; and by keeping the shape of the curves constantly in the mind, they provide mental pictures of the action of the dynamo, and enable one to remember the laws they illustrate so effectively.

(c) *Compound Wound.*—If a lamp be switched on in a parallel system, using a shunt machine, the current will increase, but the pressure will fall. If the same be done with a series machine the current will increase, but the pressure will rise. Neither will do alone, because with lamps in parallel it is necessary that the pressure should remain constant when lamps are switched off and on in order that the rest should not be affected ; but since one machine produces a rise and the other produces a fall of pressure, it is possible, by a suitable combination of the two, to produce a machine which has a constant pressure irrespective of any alteration of the number of lamps that may be burning. Such a machine is a compound-wound dynamo, which gives a constant pressure, with a varying current dependent on the number of lamps burning. "Compound wound" signifies the use of both series and shunt coils on the field magnets. This method is shown in Diagram (3), Fig. 23, where the fine shunt coil is denoted by s, and the thick series coil is denoted by F. The shunt coils consist of a large number of turns, while the series coils have only a very few.

On plotting ohms for horizontal values, expressive of the resistance of the external circuit, it will be found that the highest point of the series curve is when the resistance is lowest, and this is as it should be. The shunt curve, on the other hand, is at its lowest; following the course of the two curves, as the external resistance is increased, so the magnetising force due to the series coils diminishes, whilst that due to the shunt coils increases. Then comes a point where they cross each other. When the shunt gets near to its maximum values, and the series near to its minimum, the two curves increase and decrease respectively at a very slow rate. The ratio of the number of turns of the shunt to the series must be such that as the one increases in magnetising power the other will decrease in strict proportion, consequently the resultant effect of the two coils is to maintain a constant pressure. This condition is attained when the sum of the heights of the two curves at every point is constant, and so gives a straight resultant curve, perfectly horizontal.

A compound dynamo can only be run at some one particular speed in order to give a constant pressure under a varying load. The reason of this is not far to seek. The higher the speed of a series machine the higher the E.M.F., since E.M.F. is proportional to speed. This is when keeping the current constant. When the speed of a shunt machine is increased, and the external current kept constant by inserting resistance, then the shunt current is greatly increased, and so the E.M.F. rises much more than what it would if the shunt or magnetising current were kept constant. It is thus seen that the effect of speed is not the same on a series machine as on a shunt, therefore if a compound dynamo be run at any other speed than what it is designed for, the two sets of coils will not regulate each other as they should do.

The field magnets of alternate-current machines cannot be "excited" by using the alternating current, since the magnets would be continually magnetised and demagnetised; it is therefore necessary to employ a continuous current, and this can be done in two ways: first, by running a small continuous-current dynamo, called an "exciter," and using it simply for magnetising the field coils of the alternate-current machine; secondly, by leading, say, some coils of the armature to a "commutator" fixed on the shaft in front of the "collector"—the alternating current thus obtained from these coils being commutated into continuous currents, are then used to excite the magnets. In the first case the exciter is run at a constant speed, and so the magnetic field of the alternate-current machine remains constant in strength; in the second case, the exciting current coming from the armature may vary in value, and so promote self-regulation.

Armature Winding.

Armatures can be divided into three types, as in the case of field-magnet winding: (a) drum; (b) cylinder, or long ring; (c) disc, or short ring. Each method of armature

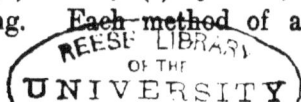

winding has its peculiar advantage, and is used for certain cases.

(a) *Drum Winding.*—A drum-wound armature is of the shape, as the name implies, of a solid cylinder, the wire passing lengthwise round the surface of the drum-shaped core. The length of wire that passes across the two end faces of the drum is evidently wasted, so far as cutting lines of force is concerned, so that the useful length of the armature wire or conductor is measured by the length of the drum ; one length of wire along the drum is called an " armature conductor." These conductors are fixed around the periphery of the drum core, diametrically opposite conductors being joined together by an " end connection " at the back end, the other two ends of the pair being connected to two contiguous segments of the commutator. The beginning of the next pair is joined on to the same segment that the end of the first pair is joined to, and so on. Therefore it follows that all the conductors are in series, so that these front connections on the commutator are similar to what is shown in Fig. 22, and by imagining the wire to be fixed lengthwise round a drum, instead of being coiled several times round an annular ring, we can obtain an idea of the connections of drum winding. To avoid the wires crossing or overlapping each other at either end, the end connections take a flat spiral form, semicircular strips of varying diameters being used for the several conductors.

Since the length of the drum is large compared with the diameter, therefore the useful length, or the conductor, is large compared with the wasteful part, represented by the end connections, back and front. The great advantage of drum winding is owing to this, because, for similar total length of wire, this method of winding will produce a greater E.M.F. than any of the other methods of winding, the waste wire being less. Drum armatures are mostly 'used for low-pressure dynamos, such as running incan--descent lamps in parallel. They are ill-adapted, however, for machines giving high pressure, because, owing to the numerous end connections, they are very difficult to wind and insulate properly, so that the risks of short-circuits taking place are great, for any two wires would have a fairly high difference of potential between them, and there is no room to provide suitable insulation. When heavy currents are required at a constant pressure, these armature conductors pass from a wire into a massive bar. The bars are sometimes made of solid copper, and sometimes built up of a number of copper strands, each strand being first insulated by varnish or any other suitable stuff. The bar is then twisted in the centre and subjected to great pressure, so as to make it as compact and solid as possible. The object of making the bar in strands like this is to kill the "eddy currents," which would circulate in a solid mass of copper. The direction in which eddy currents circulate is at right angles to the true or main current, so that by dividing the bar lengthwise, their circuit is broken. Some makers build their armature bars up by putting insulated strips together. Bars are insulated by cotton, tape, varnish, etc., and often separated by air-gaps; this latter plan has great merit, for there is no better insulator than dry air, and the ventilation is excellent, because the bars are exposed to the air. End connections are mostly soldered on, but sometimes, particularly in Continental practice, they are screwed on. This latter plan is of immense advantage when the bars have to be taken out or in any way interfered with, but it can hardly be called as safe and sound a job as soldering. If the armature core were made of a solid mass of iron, the same trouble with eddy currents would arise as would with the bars, and to a very much greater extent, so that the core must be built up with a number of thin discs of very soft iron. Each disc is varnished over to insulate it from its neighbours, and the whole compressed and held in position by two stout end plates or washers, screwed on the spindle of the core. The armature bars are driven round by being let into slots cut lengthwise in the core, and are securely attached, so as to resist the great force to which they are subjected, since when the armature is revolving, the magnetic field strives to drag away the bars. They are kept in position along their length by winding steel wire round them and the core, a layer of mica being put between the bar and the binding wire to insulate the two. The binding is put on in belts, or bands, each having from 10 to 30 turns, and being a distance of 1in. to 2in. apart, according to the length of the armature; the binding wire is then wiped over with solder. This binding wire naturally must be very thin, so as not to take up any room; as a rule there is only about a quarter of a centimetre air-space left between the binding wire and the pole-pieces, between which the armature revolves.

(b) *Cylinder, or Long Ring, Winding.*—For this method of winding a broad ring is required, the winding being round the annular body of the ring, hence as much wire passes through the inside of the ring as passes outside, and all the wire passing inside the ring is not only useless, but highly detrimental, because it produces a counter E.M.F. through cutting the lines of force that leak across the ring. The armature core is built up with a number of thin soft iron discs of a ring shape, the core being driven by a gunmetal spider, which is keyed to the shaft. This is very different to a drum core, this latter consisting of discs forming a solid cylinder, through whose centre the shaft passes, while the former constitutes a hollow cylinder, having considerable annular depth, and free from iron in its interior, not reckoning the driving shaft, from which it is separated by a considerable air-space. These armatures are suitable for high-pressure machines, such as arc lighting dynamos. They are easy of construction, easy to repair, and can be well insulated, but their efficiency is low compared to the drum.

(c) *Disc, or Short Ring, Winding.*—This kind of winding is mostly met with in alternate-current machines, where the magnets are arranged in two circles, the disc armature revolving between ; by this arrangement the opposing poles are brought very near to each other on account of the thinness of the disc. Another arrangement is to have the armature mounted so as to revolve within a circle of magnet poles that project radially inwards from a circular yoke, in which case the pole-pieces are made alternately north and south. The core takes the form of a kind of wheel, the armature wire being wound on bobbins placed round the periphery of the wheel core, the number of bobbins corresponding with the number of north or south poles. Instead of bobbins filled with wire, the winding may be in the form of copper ribbon wound on " formers." The two ends of the winding wire are led to two collecting rings on the shaft.

Closed Coils and Open Coils.—So far, only armatures of the closed-coil type have been mentioned, such as is shown in Fig. 22. These are either of the drum or ring wound pattern, and are found in shunt and compound wound dynamos, suitable for low and constant pressure work, such as running incandescent lamps in parallel. Alternating-current machines giving high pressures also have the coils joined up in series, so as to make one continuous circuit or closed coil.

An open-coil armature signifies one in which each coil or set of coils is separately connected to two diametrically opposite segments of the commutator, so that when the two brushes are resting on these two segments they collect current only from that one coil or set of coils, all the others being out of action ; this being so, it is evident that the brushes must collect the current from the coil when it is in its most active position, and it is further evident that this position is when the coil is lying parallel with the magnetic field or in a horizontal position — therefore the brushes of an open-coil armature must be at the ends of a horizontal diameter, or at right angles to what they would be in a closed-coil armature. There is no necessity, however, to restrict the collection of the current to one set of coils, because those coils which lie next to the horizontal-placed coil are cutting the magnetic field almost at as great a rate, and this can be utilised with advantage; it is customary, therefore, to allow the brushes to make contact with a pair of coils at the same time, when this pair is passing through the most active position, one set just coming into the position of maximum action, and the other set just going out. Open-coil armatures are used in series-wound dynamos giving continuous currents. These dynamos are made for high pressure and constant current, such as for running arc lamps in series with a pressure of, perhaps, 2,000 volts, and a current of 10 amperes.

Owing to the high pressure generated by the machine, and the fact that the current is collected at the point of maximum activity of the coils, it is impossible to use a commutator built up with mica-insulated segments such as is found in low-pressure dynamos, because the terrible sparking would quickly burn up and ruin the insulation and the segments themselves. It is necessary then to have the segments, which are few in number, insulated from each other by air-gaps. One great disadvantage of the open-coil armature is that the current is somewhat discontinuous, owing to the very small number of coils employed on the armature.

A closed-coil armature may have 50 conductors, each half of them being in series, and the two halves then put in parallel through the brushes, as already explained, but in an open-coil armature there may not be more than four or eight coils. The open-coil armatures may be built in a drum, ring, or disc form, but those of the ring form are more usual.

The above only gives a very brief outline of the classification and uses of the various kinds of armatures, and therefore only a few of the most important points have been dealt with.

Working in Parallel.

In central generating stations it is often essential to be able to throw dynamos in and out of circuit, according to whether the load is heavy or light. By the word "load" is signified the electrical output or energy that is delivered to the mains, and thence distributed throughout the premises or district supplied with light or power.

As may easily be understood, the load will fluctuate up and down very irregularly during the 24 hours of the day, and naturally will be high during the evening and almost nothing during daylight. Every lamp that is switched on means an increase in the electrical output, and hence increased work for the engines and boilers. Similar to distributing mains and pipes for gas, only one network of distributing mains for electricity is used for parallel working, so that of the dynamos which generate the current each must send its own share into the one circuit or set of mains. When dynamos work together, coupled up, to supply energy to a constant pressure .circuit, they are said to be "running in parallel." As the load increases so machines are started and put in circuit, and as the load decreases so they are switched out of circuit and shut down.

We will now examine the way in which this operation is done, the proceeding being what actually takes place in a central station day by day.

Let there be a battery of six continuous-current dynamos,. compound wound, yielding a constant pressure of 100 volts in the lamp circuit, and capable of generating an output of, say, 50 kilowatts, thus giving a maximum current of 500 amperes each. It is usual with this kind of machine to have a small variable resistance coil inserted in the shunt coils : the object of this is to slightly vary the resistance of the shunt, whereby the pressure of the machine can be increased or decreased a little. When the machine is working with a heavy load there will be greater loss of pressure along the mains and distributing wires than when working at light load, because the current will be greater. (The loss of pressure is given by multiplying the current in amperes by the resistance of the mains in ohms.) So that in order to maintain the same pressure at the lamp terminals, it is necessary to slightly raise the pressure of the machine as it takes the load on, by utilising this variable resistance ; but it is not used so much when a number of machines are working together as when one is by itself, because what would signify a large variation in one machine becomes only a small variation to a number. Fig. 26 gives the diagrammatic sketch of how the above-named six machines are connected up in parallel ; only two are shown, the rest being similarly connected. A A signifies the two armatures ; B B, positive brushes ; T T, positive terminals ; S S, positive switches ; B' B', negative brushes ; T' T', negative terminals ; S' S', negative switches ; P P, binding-posts for series and shunt coils. The two ends of the equaliser are shown in the diagram as connected to B' B', in order to make the diagram simpler ; in practice they are really attached to the binding-posts, P P, which is, of course, the same thing—P thus joins together one end of the series coils, one end of the shunt coils, and the negative brush cable. Sometimes each machine has a third switch, called the equalising switch, for the purpose of disconnecting the negative brush from the other negative brushes ; in that case the equaliser is brought up to the switchboard, where it is fixed below the negative omnibus bar. Each machine is supplied with positive and negative switches, all the positive terminals of the dynamos communicating by means of the positive switches to one massive copper bar, called an omnibus bar, from which the positive mains of the several circuits branch off. The positive dynamo terminal, as will be seen in the diagram, forms

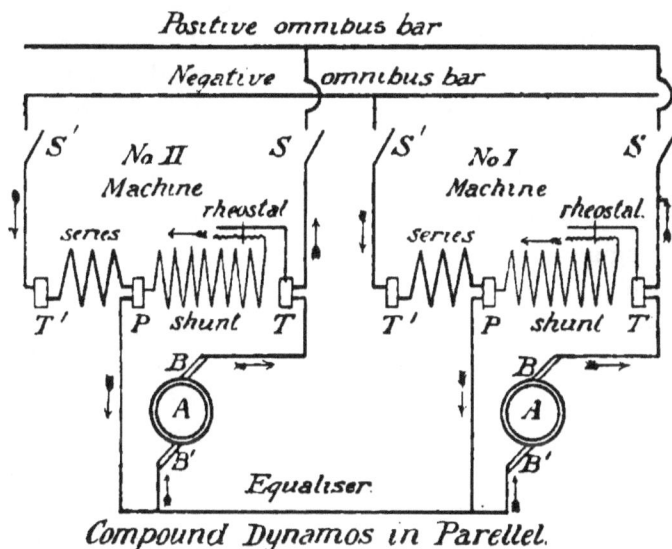

Compound Dynamos in Parellel.

FIG. 26.

the positive end of the shunt coil and the positive brush of the dynamo. All the negative terminals of the dynamos, which indicate the ends of the series coils of the dynamos, are connected by means of the negative switches to a second massive copper bar, from which the negative mains of the several circuits branch off. The shunt coils are now put all in parallel by connecting up all the negative brushes of the dynamos, which represent the negative ends of the shunt coils, to a thick piece of cable called the "equaliser," which generally is run in a chase in the dynamo-room along the line of dynamos. The current from one machine passes by way of the positive brush and positive terminal into the lamp circuit, through the positive switch ; returning through the negative switch, it passes by way of the negative terminal through the series coils, then by the negative brush back into the armature again, thus completing the circuit, a small portion of the current, of course, flowing through the shunt coils. The pressure of the lamp circuit is given by a voltmeter whose terminals are placed across the omnibus bars, and a second voltmeter can be used to give the pressure of any individual machine by touching the voltmeter key belonging to that machine.

Everything being ready, the following is the method usually adopted to put a machine in parallel : Start the engine, then put down the dynamo brushes, and turn on the pilot lamp. The machine will take a little time to excite itself, and probably a minute will elapse before the pilot lamp grows to full candle-power ; when it is thought to be about right pressure touch the voltmeter key of the machine. If too low, speed up a trifle, and stop when the pressure of the machine voltmeter is two volts higher than what the circuit voltmeter reads. Previously to this, slightly lower the pressure of the lamp circuit by, say, one volt, because when the new machine is switched into parallel and takes the load, the other machines, being relieved of some of their loads, would have their pressure slightly raised. This need not be done, however, when there are more than two dynamos working, and another one is about to be put into parallel. Now throw in the negative switch first and then the positive switch. If the engine is worked from its throttle-valve, it will be seen that although in parallel the ammeter shows no load, so that by opening the throttle-valve and giving more steam to the engine the dynamo then takes its load, the load on the other machines being then levelled down until all are as near equal as is possible.

It is most important to throw in the negative switch before the positive, for this reason. The polarity of the machine might have become reversed when last shut down, and by putting in the negative switch first, the only effect is to send an opposing current from the lamp circuit round the series coils in the same direction as flows through the series coils of the other machines, thus tending to restore, if necessary, the magnetic field to its right polarity ; whilst if the positive switch be put in first, the machine will be put in series with the others instead of in parallel, and the armature will cause a short-circuit. Upon examining Fig. 26 this action will be made clear. By the use of the negative switch another advantage is found. Suppose the load increases very suddenly, due to a fog, etc., and the machine must be put into parallel without loss of time. By throwing in the negative switch, the field is excited almost at once by a current from the circuit, hence its pressure quickly rises and prepares it for putting into parallel. This "short cut" is not to be recommended, however ; if possible, better be sure, though slow, for if in the excitement of the moment the positive switch got put in by mistake instead of the negative, and the brushes happened to be down, then—— ! ! !

Now coming to shutting down a machine. The first thing to do is to reduce the current or load of the machine which is to be shut down ; this also can be done very effectively by manipulating the throttle-valve. It is not wise to reduce the current to nothing, because in large-range ammeters, reading many hundreds of amperes, very small amounts cannot be relied upon, particularly if the instrument is a little out of adjustment, as sometimes happens, and there is a risk of the dynamo receiving a back current, so that a margin of a small current should be left. Then the pressure of the lamp circuit should be slightly raised, if deemed advisable, in order to allow of the remaining machines to take up the small load which is thrown off the machine when it is shut down. All being now ready, the driver with his hand on the starting valve, the signal is given, and the needle of the ammeter watched. Immediately it begins to move towards zero the positive switch is thrown out, thus throwing the machine out of the circuit. With a little practice, and knowing your driver, this can be done neatly at the right moment and without the smallest spark. The negative switch is thrown out after the positive.

To run compound machines successfully and efficiently in parallel, depends a great deal on the resistances of the series coils of the machines, and the resistance of the "equalising" or "balancing" wire. When the series coils of the machines have equal resistances, and when the resistance of the equaliser is low, then whatever the difference may be between the E.M.F. of each machine, the current in the series coils will be approximately equal and independent also of the armature resistances. Hence the machines should be of same size, or, at all events, should have *equal* resistance in their series turns; also the balancing wire or equaliser should be, as stated above, a thick cable having low resistance, and not a fine wire, as is sometimes stated erroneously in text-books.

It is quite as easy to work alternate-current machines in parallel, provided you know how to do it, and it is only comparatively lately that the practice of doing so has been adopted. When dealing with continuous currents, there is only the matter of pressure to trouble about, but with alternating currents the pressure and also the current rises and falls, and it is necessary for each machine to be exactly in phase at the moment when they are put in parallel—that is to say, the pressure waves of the machines must coincide and rise and fall simultaneously : this condition is what is technically called being " in step."

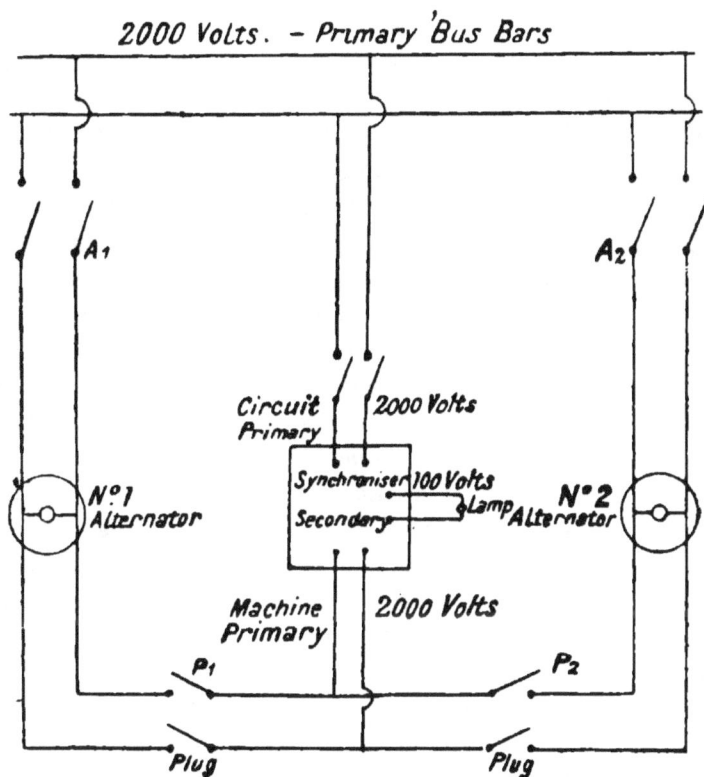

2000 Volts. - Primary Bus Bars

A_1 A_2

Circuit Primary   2000 Volts

N°1 Alternator   Synchroniser 100 Volts   Lamp   N°2 Alternator   Secondary

Machine Primary   2000 Volts

P_1 P_2

Plug   Plug

Fig. 27.

Alternators in Parallel.—To run alternators in parallel they must be connected up to a pair of omnibus bars, similar to the way low-pressure dynamos are, so that the primary high-pressure wires from the alternators are all joined up in parallel. The diagram, Fig. 27, shows two alternators in

o

parallel—No. 1 being connected by switch A_1, and No. 2 being connected by switch A_2 to the 'bus bars. Various high-pressure primary circuits are led off, each being con nected with the primary coils of a large transformer, or to a bank of transformers, according to the nature of the system of distribution ; these transformers being placed at distributing centres, whence the secondaries supply the respective surrounding areas.

But putting alternating-current machines into parallel is a very different performance to coupling up continuous-current machines, because it is necessary that the machine about to be put on the circuit should be exactly in step, or in phase, with the others. To effect this, an instrument named a "synchroniser" must be used. This apparatus is shown in the accompanying diagram ; it consists virtually of three coils, and may be called a double transformer, because it has two primary coils and one secondary coil placed between the two primary. One primary is connected through a switch to the circuit, and is named "circuit primary," the other primary is connected to the terminals of the machine that is about to be put in parallel, so that plug P_1 will connect it to alternator No. 1, and plug P_2 will connect it to alternator No. 2. The ends of the secondary coil are led to a single incandescent lamp, say, 16 c.p. We will now describe the usual method of coupling up two machines. Suppose No. 1 alternator is running on the circuit and it is desired to add No. 2 alternator, after speeding up the machine, first close the switch connecting the circuit primary of the synchroniser to the circuit or 'bus bars ; second, insert plug P_2, thus connecting No. 2 machine to the other primary of the synchroniser. Two primaries will now act inductively on the one secondary coil, the current in each primary being of different phase. The result is that, although the secondary lights up the lamp, the light will fluctuate violently, going in and out in a spasmodic manner; in a few seconds, however, the light pulsations will become more regular, and their amplitude will rapidly decrease, showing that the two machines are getting into step, and that their current waves are becoming synchronised. A few seconds more, and a moment arrives when the light of the lamp becomes momentarily steady at full candle-power ; this is the critical moment at which the machine should be switched into circuit by closing switch A_2, since the machines are then exactly in step, and their current waves rise and fall simultaneously in the two primary coils of the synchroniser. It is necessary to have some practice before the operator can seize upon the exact right moment to switch in ; he must fix the lamp with his eye, so to speak, and have his hand on the switch ready to throw in the switch at once. The whole operation should not take more than one minute, and if the opportunity of switching in be lost, another must be waited for ; a voltmeter is more reliable than a lamp.

Notes on Running

The weakest part of a dynamo, and the only one that really requires any watching, is the commutator and its attendant brushes. The field magnets, being built of solid slabs of iron, suffer no depreciation, and will last until the machine is obsolete. Almost as much can be said for the coils they carry, since they are stationary, and not subjected to any strain. With machines that are not of sound workmanship, the insulation will become defective, but this is not a serious matter, and can be repaired by the dynamo attendant. Should a short circuit occur, however, and burn or otherwise damage the inside coils, then it is a long and troublesome task to unwind, repair, and rewind, but excepting an unusual case of this sort there is little or nothing to disturb their life and soundness.

Turning our attention to the armature, this is much more

liable to accidents and damage. When a dynamo gets short-circuited, the armature may get burnt up. The damage done varies: sometimes the armature is completely ruined, at other times it is one or more coils that are damaged, when they must be taken out and repaired, or replaced. Another source of weakness is the binding wire of the armature. This is liable to break or get torn off to the extent of two or three turns.

Foremost in point of vulnerability comes the commutator, which, as before mentioned, is the weakest part, and it is the only part subjected to wear and tear. Here, the brushes are constantly rubbing and wearing, and this, together with a more or less amount of sparking, will very rapidly make its effect apparent, unless great care and attention is exercised. A dynamo attendant takes as much pride in the condition of the commutator and the brushes as an engine-driver does in the smooth running of his engine; and so he ought to, since it is the only part of the dynamo that requires any attention. A well cared for and preserved commutator presents a highly burnished darkened appearance, is perfectly true and smooth, and has no scoring on it. The state of the commutator depends almost entirely on the state of the brushes and the way in which these latter are adjusted.

Dynamo brushes are of various types. The most common form in use for incandescent dynamos, with the closed-coil armature, is that known as a gauze brush; this brush is made up of a length of copper gauze folded round several times, so as to make a solid thick strip, about three times as long as it is broad, and from a $\frac{1}{4}$in. to $\frac{1}{2}$in. thick, the size of the brush being increased as the machine becomes larger. This kind of brush being of a soft and yielding nature, makes a springy contact on the commutator, and so to some degree diminishes cutting or scoring, but against this there is the disadvantage of the brush wearing more quickly. Commutators are made of hard rolled copper, gunmetal, or phosphor bronze, but it is naturally better to have the brushes wear than the commutator, for the former can be easily and cheaply replaced while the latter cannot. Another good kind of brush, very suitable for small dynamos, is formed by a number of thin brass strips laying one on top of the other, each strip being cut, say, $\frac{1}{16}$in. shorter than the one on top of it; being very flexible, the several ends make contact on the commutator when the brush is placed flat or tangentially on the surface. Brass being harder than copper gauze, does not wear so quickly. The great advantage of a brass-strip brush is that the trimming is next to nothing. Every three or four days—according to the number of hours the machine works—the tips feather out, and all that is necessary is to clip the tips off with a pair of shears, or strong scissors, and then push the brushes forward a trifle in the brush-holders when replacing them, so as to ensure their occupying the same position. Another advantage of this kind is that they can be home-made, and at a very small expense; when the various strips are cut of a suitable length and width, the bottom $\frac{1}{2}$in. must be well soldered, in order to hold the strips together. Prior to the gauze brush the most common form was a bundle of fine copper wire, one end being soldered, while a metal band was slipped over the other end to keep the wires in place, the dimensions being similar to the gauze brush, but it cuts the commutator more than the gauze does, and is more troublesome to trim.

For arc lamp dynamos it is usual to use a brush made of a fairly stout strip of brass, about one-sixteenth of an inch thick and, say, 8in. long, the width being from 2in. to 3in., according to the size of the machine. One end of this strip is split into about five or six tongues, extending a suitable way along the length, and the split end, being thus made flexible, presses on the commutator. For alternate-current machines, light gauze brushes are suitable, the current being only small.

Carbon brushes are found to be of great service when the load on a dynamo varies greatly, because with a fluctuating load sparking is very prevalent, and carbon reduces this to a great extent; for this reason, they are particularly suitable for motors, and in America scarcely any other kind are used for traction work. They are fixed "butt" end on the commutator, because a motor armature is required to revolve in either direction, and the holder is fixed quite close to the commutator in order to form a support to the brush end, carbon being somewhat brittle. For equal section they will not carry so much current as metal will, therefore 50 per cent more cross-section should be allowed; having them copper plated gives a better electrical contact in the holder.

A new kind of brush lately introduced is that known as the "foliated" brush (Boudreaux patent). This brush is greatly used on the continent, and is also gaining favour in England. It is made in a block consisting of a great number of very thin copper leaves (1,000 to the inch); the result is that it.has an extremely soft rubbing surface, perfectly smooth. Owing to the small amount of antifrictional metal that is present, it cannot wear away the commutator, and lasts, itself, a long time. The author has tried these brushes, and found them to keep the commutator in a beautiful polished state.

We have stated what the appearance of a well-preserved commutator is like, and will follow this up by giving hints and advice how to keep it in that state. The effect of a brush pressing and rubbing on a commutator tends to produce a hollow along its path. This hollow is produced by three main causes—first, length of time; second, pressure of the brush; third, hardness of the brush. Then there is the detrimental effect of sparking, which burns away the segments and the brushes.

However well the brushes are adjusted, the constant friction of the brush is bound in time to wear away a groove. The result is that between each brush there appears a slight ridge. These hollow ridges can be felt before they are seen by passing the finger lightly across the commutator. The merest indication of a ridge or path should be promptly combated, by treating it with emery-paper. If they are allowed to increase, then emery-paper will no longer be of use, and a file must be put on, and if neglected beyond this the hollows will develop into "ruts," and nothing short of having the commutator turned up will make a good job of it. The evil of these hollowed paths is that the brushes are worn unevenly, and cannot be adjusted ; result is great sparking, and this, with the bad state of the brush, increases the bad state of the commutator. But since "prevention is better than cure," means should be taken to prevent even the smallest ridging, and the only way to do this is to set the brushes in such a position that the positive ones lie on the commutator, covering the gaps between the negative ones. A brushholder being, say, $\frac{1}{4}$in. to $\frac{1}{2}$in. wider than the brush it carries, there is a gap between the brushes when two holders are set close together ; but with the above arrangement, all the surface of the commutator between the two outside brushes gets equally worn. A great fault of some commutators is that they are too short to permit of this, there being just room to take in a set when placed exactly opposite each other, and no more, or at the most not enough to be of any use.

We now reach the second cause—viz., pressure of the brush. This does not as a rule receive the attention it should do, particularly when dynamos are to a great degree left to run themselves, as is often the case in mills and factories ; or else are looked after by those who know very little about them. The tension of the springs should be just sufficient to cause the brushes to make a light, yet reliable, contact with the commutator. The contact must not be too light, because the brushes will vibrate and so cause sparking, and if they press too hard they will wear away both themselves and the commutator to an undesirable extent, and in addition will become unduly heated owing to the great friction. The best proof of the grinding pressure of a brush is the amount of brass or copper dust thrown off: the upper side of the negative or bottom brushes will become covered with this dust, as also will the holders. The tension of the springs of the bottom brushes should be made slightly greater than that of the top brushes on account of the weight of the former. An idea of the proper pressure can be obtained when the brushes can be pushed off the commutator by the little finger with ease.

Reaching the third cause, that of hardness of brushes, it is evident that a brass-strip brush will wear away the commutator more than a soft copper-gauze brush will do; something must wear—in the former case it is the commutator, in the latter case it is the brushes, so that the harder the brushes are, the less should the tension of their springs be, if this is possible. To prevent cutting of the commutator by these hard brushes a lubricating matter is often put on the commutator—the best vasoline is the only lubricant that should be used; when used, only the very smallest amount should be smeared on a finger tip, and then applied by passing the finger across the commutator. It reduces wear and tear by decreasing the friction, and gives a good burnished appearance to the commutator. It will be found, however, that on account of the mica insulation being harder than the copper segments, the latter, in wearing away, will leave the former projecting slightly. This will cause the brushes to kick, and so promote sparking; therefore it is necessary at times to apply fine emery-cloth, running the engine at full speed. These projections can easily be detected by the finger nail. Some people smother their commutator in bearing oil, with the result that the brushes get clogged up with carbonised oil and dirt, and trouble is sure to follow such practice, for this matter forms a film over the segments and tends to short-circuit the armature coils. This last objection does not apply to alternate-current machinery, nor to arc light dynamos, because the first have ring collectors instead of commutators, and the brush never breaks contact, and the second have air-gaps between the segments. The whole of the brush-carrying gear should be kept scrupulously clean, and once every week or so, according to the number of running hours, the gear should be taken entirely to pieces, by taking off the carrying bars, and cleaning the holder parts, spring action, etc. Sometimes the carrying bar will be partly covered by a thin black film of oil and dirt, thus obstructing proper electrical contact between the holders and the bar. This and another point to which particular attention is drawn, is that the insulating ebonite or vulcanite washers which insulate the carrying bar from the rocker should be well cleaned. Through being near the end bearing, with defective or poor oiling arrangements, oil sometimes creeps or gets thrown off, and settles in a very fine film over the surface of the rocker, and also over the insulating washers, the proof of which is that these become sticky ; and should the washers get sprinkled over with copper dust thrown off by the brushes, a film of oil and dust is formed having the property of a partial conductor. These washers are only about $\frac{1}{8}$in. thick, and are all that insulate the positive brush from the negative, for the rocker is made of iron, and if the insulating qualities of the washers deteriorate or break down, there is a dead short-circuit of the armature.

The most suitable angle at which brushes should make contact with the commutator is about 45deg., and most of these kinds of brushes are therefore shaped to this angle at their contact end. The tip of the back of the brush is called the "toe," and the beginning of the angular contact face at the front of the brush is called the "heel."

The following figures are exact measurements taken from various-sized continuous-current machines for incandescent lighting : The smallest is an eight-kilowatt dynamo having an output of 80 amperes at 100 volts. There are four brushes, two being positive and two negative, the dimensions being 18cm. long by 5cm. wide by ·3cm. thick ; the width of the segments of the commutators is ·5cm.; the width of the insulating strips between the segments is ·1cm. These brushes being only thin, they make contact at an angle of 60deg. instead of 45deg.; calculated out, the contact surface of each brush comes to about 3 square centimetres; each brush carries at full load 40 amperes, and the current density in the brush works out to about 26 amperes per square centimetre. For a larger machine of 44 kilowatts, having an output of 400 amperes at 110 volts, there are six brushes in all, whose dimensions in centimetres are 18 × 5·3 × ·8 ; the segments are 1·5cm., the insulating strips ·1cm., the contact surface of each brush is about 6 square centimetres, and at full load carries 133 amperes, and has a current density of 31 amperes. The largest machine, of 126 kilowatts, has an output of 1,150 amperes at 110 volts. Dimensions of brushes in centimetres are 18 × 5·3 × 1·1 ; 12 brushes are provided ; the segments are 2·8cm ; the insulating strips ·15cm.; the contact surface of each brush is 9 square centimetres, and at full load carries 190 amperes, hence its current density is 32·5 amperes per square centimetre.

Gauze or wire brushes require to be taken out of their

holders and examined every other day or so, and, as a rule, it will be found that they will run 20 to 30 hours before they require retrimming. The contact face becomes worn to a slight curve concentric with the commutator, and when oil is used on the commutator, or when it is in a dirty condition, the brush face will have a black slimy coat of carbonised oil and grit ; in any case, the face when dirty must be rubbed over with a round file and brightened up. When the face is worn or burnt away unevenly, it must be filed up true, and for this purpose it is well to have a wooden block in which to fix the brush; this block consists of two pieces, both shaped at one end to 45deg., the back part has a groove cut in it just wide and deep enough to receive the brush, while the front part covers the brush, which is thus enclosed in the block. The block holding the brush is put in a vice, and the face of the brush can then be filed up perfectly true and to the right angle. When the brushes are well and regularly looked after, they very seldom require shaping with the file. All that is necessary is to clip off the fringed ends and corners with a sharp pair of scissors or shears. It is the toe of a brush that requires clipping, and it is a very good test with gauze brushes to hold them up to the light with their back towards you ; if the toe is feathered out so that there is only one layer of gauze at the tip, or if it can be clearly seen through, then cut off this part. On replacing the brushes in their holders, the greatest care must be taken to see that the contact face makes perfect contact. If the brush be pushed too far forward it will touch at the heel and not at the toe and if it is not forward enough it will touch at the toe and not at the heel. The brush must also be put in perfectly straight, so that it lies at right angles to the length of the commutator. A light should be provided when adjusting the brushes, so that by placing it just under them the contacts can be clearly seen and scrutinised.

When adjusting the negative brushes, which are usually underneath the commutators, the negative brush cable can sometimes easily be disconnected, also the positive, and the rocker can then be swung right round, thus bringing the negative brushes into the place of the positive, where they can be adjusted in a satisfactory manner, and with more personal comfort. Sheet-metal brushes require scarcely any attention: simply clip off the feathered ends and wipe them with a linen rag.

Whether driving by belt or direct, always start the engine before putting the brushes down on the commutator, and never put the full load on the dynamo suddenly or altogether; have the circuit switches open when the main switches are put on, and then, as the machine takes the load, the several circuits may be switched in. Owing to the magnetising effect of the armature coils, the armature soft iron core becomes an electromagnet. The result is that the two magnetic fields belonging respectively to the field magnets and the armature core react on each other; but the former being very much stronger than the latter, the resultant field lies a little out of the normal position, the lines of force being shifted round a small angle in the same direction as the direction of rotation of the armature, and since the diameter of commutation must be at right angles to the resultant field, hence the brushes must be shifted forward likewise. The heavier the load the more magnetised becomes the armature core (within certain limits), consequently the more must the brushes be shifted forward. This movement forward is technically called giving a "lead" to the brushes; for full load the lead does not amount to more than, say, 15deg. beyond the vertical line, but this varies a good deal in various machines The proper amount can only be determined by practice, and the best lead is found when the brushes are moved to a non-sparking position. As the load varies on a machine it is necessary for the dynamo attendant to shift the brushes slightly by giving them an increase of lead when the load increases, and a decrease of lead when the load decreases.

When shutting down a machine, reduce the load gradually by easing down the engine, then when the lamps are almost out, or a very small current is flowing, the circuit or circuits, if several, may be opened ; this prevents heavy sparking at the switches and the tendency for the engine to race. Before stopping, lift the brushes off the commutator, otherwise the brushes will be damaged in the event of the engine making a backward motion, which it often does. A caution may here be given respecting large shunt machines. After slowing down the engine, wait two or three minutes before attempting to lift the brushes off the commutator. The reason is that it *takes time* for the current to die out of a high-resistance coil, just as it takes time for the current to grow to its full value when starting a current in a circuit. Some machines are supplied with a switch, which puts a resistance in parallel before breaking the shunt. The circuit of the shunt coils is completed through the armature, hence if it be broken before the current has decreased to zero, the self-induction of the coil will produce a disruptive spark, and probably injure the insulation of the coils. When the brushes have been lifted up, the commutator must be rubbed with emery-cloth if it requires it, a fairly quick speed being maintained ; this polishes up and smoothens the commutator. After the emery-cloth, a clean duster must be applied. Waste should never be employed for cleaning or polishing the commutator, brush-holder, or any such parts, neither should fluffy cloth ; linen only should be used, because this does not leave any threads or fluff behind, and it wipes clean. The brushes should then be inspected, and trimmed and adjusted, if necessary. It is advisable to get all this done at once, after the machine is stopped, so that the dynamo is perfectly ready to be re-started at a moment's notice, for in central stations there is no knowing when a machine may be wanted—the load may run up, due to a fog, thunderstorm, etc., or anything occurring out of the usual, such as breakdown of a machine, etc. The commutator should be provided with a suitable covering to protect it when the machine is shut down. Lastly, all connecting wires should be examined, to see that they are firmly fixed, because terminal screw-bolts sometimes work loose owing to vibration, especially in shiplighting plants; shunt coils and other fine wires leading to rheostats or regulating gear, should be particularly examined, and also occasionally during running—these fine wires are not always so well fixed as the larger ones. The connecting screws being small will not stand much, and they seem to get loose more often than the others.

Sparking at the brushes arises from several causes, chiefly, however, to neglected state of brushes and commutator, such as bad trimming, bad adjustment, rough commutator, etc. Some arise from the way in which the dynamo is made, and these, of course, cannot be rectified without altering the construction of the machine. The most likely causes are classed as below :

(1) Badly-designed dynamo.

(2) A badly-wound armature, giving an irregular rise and fall of E.M.F. round the commutator.

(3) Defective insulation in the field-magnet coils.

(4) Defective insulation in the armature coils.

(5) Defective insulation across commutator segments, caused by foreign ingredients, such as copper dust, film of carbonised oil, grit, etc., bridging across.

(6) Rough and uneven surface on the commutator.

(7) Bad condition of brushes, defective contact or adjustment.

(8) Loose connections, terminal screws, etc.

(9) Too heavy a load for the machine.

(10) Leakage of current, due to bad insulation of the circuit or the machine, thereby causing a heavier current to flow than what is utilised.

(11) General dirty condition of commutator, brush gear, etc.

If it becomes desirable, such as through a lengthened run, to trim the brushes whilst the machine is running, one brush may be taken out at a time. If the machine is heavily loaded, the brush should first be raised off the commutator very slowly to see if the rest will carry the current ; should violent sparking occur, the brush should be put down again.

Cost, Output, etc., of Dynamos.

In the early days of electric lighting a machine that would run, say, two or three hundred incandescent lamps was considered to be a large-sized one, but as time went on their output became greater and greater, until now, when the electric light industry may be said to be solidly fixed on sound foundations, for lighting, we have machines built for running 10,000 incandescent lamps of 16 c.p., and for traction, machines of 2,000 e.h.p., or 1,500 units. The Westinghouse Company put down 12 alternators for running the lights at the Chicago Exhibition, each capable of running 15,000 lamps of 16 c.p. ; the armatures of these enormous machines alone weighed 18¼ tons each. These alternators are believed to be the largest yet run by belting, and it may not be long before such sizes as these, and in all probability larger than these, will be used for generating electricity in

central stations. With arc lighting machines working at a
constant current, there is a limiting factor, which is very
difficult to get over, and that is the pressure. Every
additional arc lamp in the circuit signifies another 50 volts
pressure at the dynamo terminals. The largest continuous-
current arc lighters run 60 lamps in series, and therefore
they have a working pressure of 3,000 volts. It is very
difficult to insulate armatures of continuous-current machines
working above this pressure, the high pressure being ruinous
to the insulation. In addition to the inability of the armature coils to withstand such pressures, the greatest care must
be exercised by trimmers and circuit men in handling the
lamps, wires, etc. Since a full arc lamp takes about 10
amperes, and absorbs a pressure of 50 volts or so, therefore
60 lamps in series will require 3,000 volts pressure to
maintain a current of 10 amperes through the lot, so that a
60-arc lighter has an output of 30 kilowatts, or units, each
lamp taking half a unit.

The output of electric machines is reckoned in kilowatts,
often called for shortness, units, because the kilowatt is the
unit of electric power. This expresses the power of the
machine or *rate of work* done, and the word " unit " applied
as above must not be confounded with a Board of Trade
Supply unit, which is also often called a unit for shortness—
this latter signifies a kilowatt of electric power working for
the space of one hour, and producing a kilowatt-hour of
electric energy, or *quantity of work* done.

There are many commercial data concerning dynamos that
are very interesting and, what is more, extremely instructive.
Among such are : weight, floor space, speed, output, etc., of
machines compared with their first cost ; and different pattern
machines can be compared, when figures can be worked out,.
giving such data as number of watts a certain dynamo can.
develop per pound dead-weight, the machines compared being,

·of the same, or almost the same, output, speed, type, and
such matters ; then machines of greater sizes can be similarly
compared.   Another interesting matter to know is the number
of watts per square 'oot of floor space machines can give for
various sizes and speeds.   A third matter is the cost per unit
for various sizes and speeds.   For all the above, curves can
be plotted, and it is surprising what a quantity of suggestive
information can be gathered from them.

Speed and Cost.—The speed at which a dynamo runs has
a good deal to do with its price, and within certain limits—
say, 20 per cent. above the normal and 20 per cent. below—
the price may be put down roughly as indirectly proportional
to the speed ; for example, several dynamos, all having the
same output, and of the same design and type, will be built
for different speeds, and putting 1,000 revolutions as the
average number of revolutions per minute for what are
termed high-speed machines, we should have others running
at 800, 900, 1,100, 1,200.   Putting down the machine at
1,000 revolutions at £100, if we wanted one of similar
output and type to run at 900 the probable price would be
£100 × 1,000 ÷ 900 = £111—the lower the speed the
greater the price ; whilst one running at 1,100 revolutions
would probably cost £100 × 1,000 ÷ 1,100 = £90, so that
the higher the speed, the lower the price.   The reason of this
is very evident, for by lowering the speed it is necessary to
increase the dimensions of the machine, and hence the price.
Any increase beyond 1,200 is confined to very small machines.
Passing on to what may be classed as low-speed machines,
the average is about 400 revolutions per minute.   The afore-
said limits may be put down as generally somewhere near to
300 and 500, and the same relation between speed and price
holds true here.   High-speed machines are driven by belts, a
large driving wheel on the engine being used.   Assuming
·a dynamo pulley, or " driven pulley,", of 1ft. diameter, has a

P

speed of 1,000 revolutions per minute, it would require a "driving pulley" on the engine, of 5ft. diameter, when the engine has a speed of 200 revolutions per minute; and by employing intermediate shafting dynamo high speeds can be produced from very slow ones of the engine. Low-speed dynamos are used when the dynamo shaft is coupled direct to the engine driving shaft.

As a rule, the larger the output of a machine the lower is its speed; while very small dynamos are run at extremely high speed, say 1,200 to 1,500 revolutions per minute, very large machines do not run much beyond 300 or 400 revolutions per minute when direct driven, and 600 or 800 when belt driven. The true way to compare speeds of machines is to take the "surface velocity." For example, two dynamos may be running at the same number of revolutions per minute, but one may have the armature of larger diameter than the other armature, so that the larger armature is really running at a higher speed, because its surface velocity is greater. The surface velocity is obtained by multiplying the circumference of the armature in feet by number of revolutions made per minute. This is expressed by the formula

$$s = \pi d n,$$

where s = surface velocity in feet per minute, d = diameter of armature in feet, n = number of revolutions per minute, and π = 3·1416. A Siemens machine of 46 units has an armature of 1·33ft. diameter, and runs at a speed of 460 revolutions per minute. In this example we have s = 3·1416 × 1·133 × 460 = 1,921·8ft. per minute; this is for low-speed machines. As a rule, from 2,000ft. to 3,000ft. per minute are the working limits, although 5,000ft. is obtained with small high-speed dynamos.

Output and Cost.—The cost of machines is fairly proportional to their output, when the output does not fall below

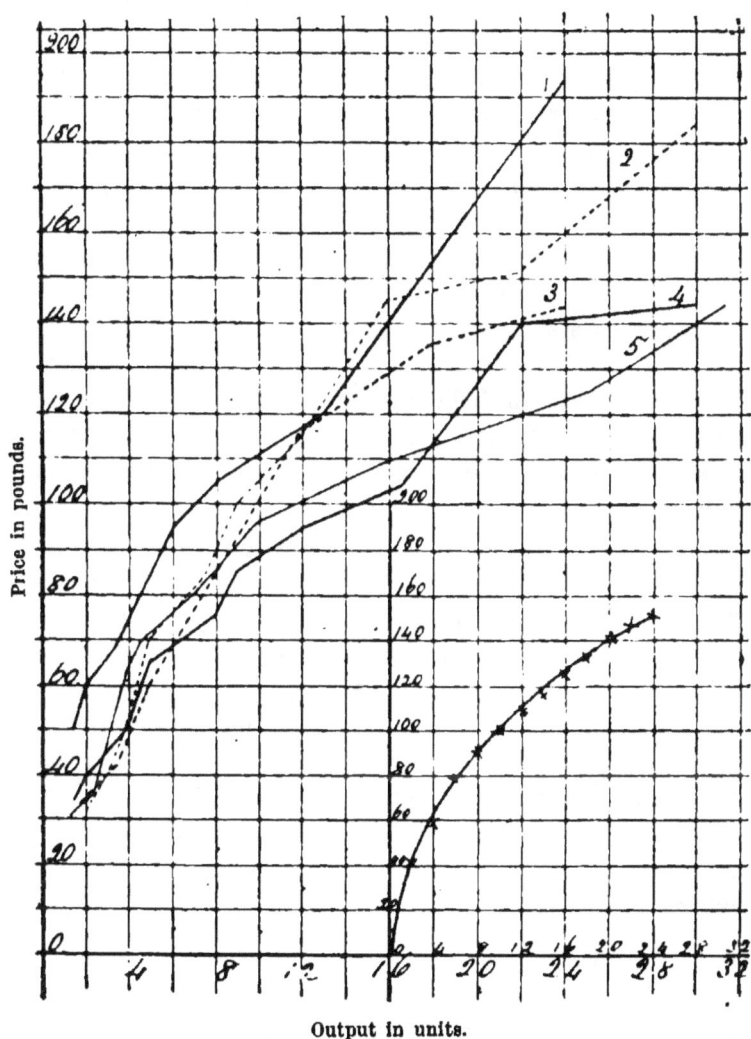

Output in units.

FIG. 28.

certain limits ; like other kinds of machinery, the cost of
dynamos per unit rapidly increases as they become smaller.
Fig. 28 gives several curves showing how the price increases

as the output increases—the curves are plotted from figures relating to different well-known dynamos. Owing to the irregularity of the curves, it was found impossible to draw curves of any pronounced character, hence they have been drawn from point to point.   It is interesting to see how they cross and recross one another, one machine being cheap at one output, but dear at another, etc.

The curve drawn at the right-hand side in Fig. 28 represents the average cost of all the dynamos for various outputs ; it is drawn on a scale of one-half that of the separate curves. The values of this averaged curve are seen to map out a distinct path of a parabolic character.   Analysing it, we note that a very small dynamo, say, of one unit output, probably the smallest kind made (except toys) would cost £24, and would be capable of running about 15 lamps of 16 c.p., or a couple of arcs of 1,000 c.p. ; taking a larger machine, one of five units, the cost comes out about £70, or £14 per unit, whilst a machine of 24 units would probably cost not more than £150, or £6. 4s. per unit.   Beyond this point the decrease of cost per unit is very slight, if anything ; and the curve will probably travel in nearly a straight line from 24 units.   All the dynamos whose curves are given have a speed close on 1,000 revolutions per minute for all outputs, thus enabling a better comparison to be made, and the following are details of the machines—the number of the machine referring to the curve :

No. I. dynamo.—Ring armature, two-pole double magnetic circuit.

No. II. dynamo.—Drum armature, single magnetic circuit.

No. III. dynamo.—Ring armature, two-pole, double magnetic circuit.

No. IV. dynamo.—Ring armature, single magnetic circuit.

No. V. dynamo.—Drum armature, single magnetic circuit.

These five machines have been selected from five districts in England : No. I. representing the Lancashire, or North-West ; No. II. the Midland ; No. III. the London ; No. IV. the North ; and No. V. the South-West.

With regard to slow-speed machines of 400 revolutions per minute, the cost would be higher than what it would be for a high-speed machine of similar output, because the slow-speed machine must be of greater dimensions and weight. A 24-unit machine would probably cost not far off £300—this is at the rate of £12. 10s. per unit; and one of 50 units about £500, or £10 per unit. It is thus seen that a machine of 400 revolutions per minute costs twice as much as a machine running at 1,000 revolutions, or two and a half times the speed.

Alternators are mostly built for heavy outputs, because they are used for distributing light over wide areas. A slow-speed alternator, running at, say, 400 revolutions per minute, and capable of developing up to 200 units, would involve an outlay of about £1,400, including the "exciter"; this is at the rate of £7 per unit. This machine would run 3,000 incandescent lamps of 16 c.p. each, thus costing a little over 9s. per lamp.

Output and Weight.—The weight of dynamos with respect to their output varies for different types and makes, because there are so many details of construction to be taken into account, such as speed of the machine, whether the magnets are made of wrought iron, mild steel, or cast iron, the density of magnetisation employed in the iron, the compactness of the machine, the bed-plate, position of centre of gravity, the current density, limiting temperatures of coils, and general features of design, such as ring or drum armature, single or double magnetic circuits, bipolar or multipolar. In addition, machines are built with modifications for certain conditions of working. From this it may be gathered that it is not wise to assume a machine is very efficient or superior to others because it can develop more watts per pound dead-weight than other machines can—the word "efficiency" being here employed to signify the life or lasting powers of the machine, the smooth running, absence of repairs, and all things that pertain to true economy. It may be put down, however, that each pound of weight will yield from five to eight watts of electrical power. Criticising machines of various outputs, the weight appears to be fairly proportional to the output— about six watts per pound is a good average ; but, as just now explained, one make may give a higher and another a lower value, so that this estimate simply gives a rough idea. The following particulars will serve as the best guide on this subject.

TABULATION 23.

Output in units.	Speed per min.	Armature.	Magnets.	Weight in cwt.	Watts per lb.
8	1,100	ring	2 pole double circuit	10·75	6·6
22	940	drum	2 pole single circuit	30·0	6·0
50	500	,,	,,	78	5·7
120	600	alternator	10 pole	123	8·7
150	350	,,	12 pole	261	5·1
180	350	drum	2 pole single circuit	265	6·0
200	500	,,	,,	340	5·2
900	—	alternator	—	1,330	6·0

All the above machines are by different makers except the second and third, which are by the same ; the fifth is an alternating-current machine, with stationary armature and revolving magnets ; the seventh machine is used as a generator for electric railway work, and therefore is more massively built than it would be if used for lighting pur- poses. This is on account of the great fluctuations of load

to which generators are subjected, thus producing great
strains; hence, power generators give a smaller figure than
lighting generators.

Output and Floor Space.—With similar pattern of dynamos
running at about the same speed, the floor space occupied in
some cases increases pretty well in proportion to the output,
but this relation is liable to be altered from numerous causes.
From two to three units per square foot may be taken as an
average estimate for dynamos above a certain size running
about 1,000 revolutions per minutes, some builders giving
higher figures, while others give lower. A few statistics will,
however, show best as when considering output and weight.
Working out figures for direct-driven machines, it will be
found that for heavy outputs the output per square foot is
high; as an example, a dynamo running at 325 revolutions
per minute gave an output of 126 units, with a floor space of
only 28 square feet, this being 4½ units per square foot.
Another direct-driven dynamo of 46 units, running at 460,
required 13½ square feet, and so gave 3·4 units per square
foot. Concerning alternators, four machines of one type,
ranging from 30 to 100 units, gave 1·4 to 1·5 units per
square foot, the speeds being from 600 to 400 ; whilst the
fifth, of 150 units, gave two units per square foot, the speed
being 350.

Probably the most interesting matter, however, will be
regarding the space occupied by "combined plants." This
term signifies where the steam-engine drives the dynamo
direct, the two shafts being coupled up together, and both
machines being on one long bed-plate. This arrangement is
sometimes named a "steam dynamo." Willans' central-valve
high-speed engines stand in the front rank, and occupy a
remarkably small floor space, probably smaller. than any
other make. Coupled up to one of Siemens's dynamos, the
over-all dimensions of one plant were 9½ft. by 4ft., or a floor

space of 38 square feet. The output of the dynamo being 46 units, this comes out to 1·2 units per square foot, the speed being 460 revolutions per minute, the engine alone occupying a space of 5ft. by 3⅓ft. Another plant, with same type and size engine, but a different make of dynamo, gave when measured a total of 10½ft. by 3⅓ft., or 35 square feet, the output being 44 units. This is 1·26 units per square foot, but here the speed is higher, being 485 revolutions per minute.

A larger-sized plant, a fine type for central-station work where ground is very valuable and floor space limited, measures 15½ft. by 4¾ft., or 73·6 square feet, the output of the dynamo being 126 units at a speed of 325, thus yielding 1·72 units per square foot.

The above will serve as good examples as to how floor space is related to output; in ship work this is a matter of paramount importance, and the same when machinery has to be fixed in the heart of a large and crowded city.

The Electromotor.

The function of an electromotor is the inverse of that of a dynamo. The latter is a machine which receives mechanical power and yields electrical power; whereas the former is a machine that receives electrical power and yields mechanical power. A most simple and beautiful illustration showing the broad principle which governs the action of the electromotor was remarked by the writer, who happened to hold an incandescent lamp near the magnets of a dynamo at work in order to examine some defect. The filament of the lamp was drawn on one side so as to bend it, thus showing the pull that magnetism exerts in a conductor carrying a current. Of course, in this case the pull was visible, due to the delicate construction of the filament and the heavy current it was carrying (compared to its sectional area). In dealing with an active copper conductor no visible movement would be observed, although it would be subjected to a magnetic pull just the same.

Any well-constructed dynamo can be run as a motor, provided the same conditions be observed : if the dynamo be a constant-current machine, then it must be run from a constant-current circuit • of same value ; and if a constant-pressure machine, it must be run from a constant-pressure circuit of same or nearly same pressure as the dynamo is built for.

At the Crystal Palace Electrical Exhibition in 1892, a number of dynamos directly coupled up to steam-engines were converted into electromotors by supplying them with current, and so were made to put the engines into motion, providing an impressive and striking lesson respecting the wonderful adaptability of electricity. Owing to this feature of " reversibility," the construction of a motor is upon lines similar to those of a dynamo ; and in most cases it would be difficult to say, without knowing, whether the machine, unmounted, was a dynamo or a motor, although, as a rule, motors are built heavier and with greater mechanical strength of parts than dynamos are.

Efficiency and Speed.—Before touching upon the various types of electromotors and their particular uses, the elementary principles which determine the efficiency and speed of motors will be explained, because the actions of a motor are rather peculiar to those who have no knowledge of this kind of machine.

The germ of the principle is the same as that governing a dynamo. In a dynamo, if a conductor be moved in a magnetic field so as to vary the lines that thread through the coil, an E.M.F. is generated in that coil, and hence a current when the circuit is completed ; so that, given two things, movement and a magnetic field, the third, a current of electricity, is produced. In a motor, however, the two things given are a current of electricity and a magnetic field, and the third is produced—namely, movement of the current-carrying conductor ; so that if a current of electricity is sent into a conductor or coil placed in a magnetic field, that coil will receive motion, and tend to set itself at right angles to the lines of force. If a number of coils symmetrically arranged in the form of an armature be inserted instead of the single coil, continuous motion would be imparted to the armature, as each coil tends to set itself at right angles to the magnetic field.

When a motor is at work it performs the double function of a dynamo and a motor, because the result of the armature of the motor being in motion induces a counter or back E.M.F. When the motor is standing still there is no back E.M.F., but it increases with the speed ; and the faster the motor runs, the greater is the back E.M.F. that is generated. The current that the motor receives must be proportional to the difference between the E.M.F. of supply and the back E.M.F. ; hence, upon starting a motor, it will absorb an enormous current, because there is no back E.M.F., and it is necessary to insert a resistance in series with the armature (and hence in series with the field coils when it is series wound) in order to reduce the current to a safe amount, otherwise it would burn up the armature. Putting E for the supply pressure and e for the back pressure generated by the armature of the motor, then the difference is $E - e$, and if R represents the total internal resistance of the motor, then the current that will flow will be $\dfrac{E - e}{R} = C$. To calculate the efficiency we must compare the electrical power or work put into the machine with the work given out by the machine—

the power absorbed $= C\,E$;
the power given out $= C\,e$;

hence the efficiency is $\dfrac{C\,e}{C\,E} = \dfrac{e}{E}$;

and is measured by the ratio between the pressure supplied and the back pressure generated. It is evident that when the motor is running as fast as it can possibly go, that will be when there is no load on it, and at the same time it will then absorb a minimum of power, because the back E.M.F. will then be at a maximum, and so the current flowing will be a minimum. On the other hand, when the motor is running very slowly, similar to when just starting a tramcar in motion, then it will have a very heavy load on it, and will draw a maximum amount of power, because the back E.M.F. will be almost at a minimum, and consequently the current will be at a maximum. The electromotor then adapts itself in a most fortunate way to its load, and on this account it is particularly suitable for propelling tramcars and the like, as it is just at the starting moment that an exceptionally great amount of power must be developed, in order to overcome the dead-weight of the vehicle, and this is exactly what a motor will do ; because when stationary a rush of current takes place, due to there being no back E.M.F., and so provides the necessary starting torque.

The useful work done by a motor was stated to be $C\,e$, and C was stated to be $\dfrac{E - e}{R}$, therefore substituting this last expression for C, we have work done $= \dfrac{e\,(E - e)}{R}$.

It is very easy to calculate the pull on the armature of any motor—only three things are required to be known. These are :

1. The density of the field.
2. The current in the armature.
3. The length of the armature.

Let the density be in lines per square centimetre ;

„ „ current be amperes ;

„ „ length of the armature be in centimetres ;

then pull in dynes = density × current × length ÷ 10 ; the expression is divided by 10 because the current must be changed to C.G.S., or absolute units, which have a value of ten times the practical unit. Taking an example, suppose the armature of a motor carries 120 amperes, is 60cm. long, and that the field density is 11,000 lines per square centimetre ; inserting values we have—

$$\text{Pull in dynes} = \frac{11,000 \times 120 \times 60}{10} = 7,920,000.$$

A force of 981 dynes represents a weight of 1 gramme, and 454 grammes make 1lb., hence

$$\text{Pull} = \frac{7,920,000}{981 \times 454} = 18\text{lb}.$$

Putting C for current in amperes ;

L for length of armature in centimetres ; and

D for density of lines per square centimetre ;

we have   Pull in pounds $= \dfrac{C \times L \times D}{4\cdot5 \times 10^6}.$

Direct-current motors may be divided into two distinct classes, just as dynamos are divided into classes. These two are constant-pressure motors and constant-current motors.

Constant-Pressure Motor.—Motors made to run from a constant-pressure source of electric supply can be either shunt or series wound. Shunt motors have several features in distinction to series-wound motors. They will run at very nearly a constant speed, irrespective of what the load may be ; and if the field be compounded, then they will run at a perfectly regular speed. In dynamos it is necessary to give the brushes a forward " lead " in proportion to the load or number of lights supplied, but in motors of all classes the opposite holds good—that is, it is necessary to give the brushes a backward " lead "; and the greater the amount of work which a shunt-wound motor does the greater must be this back " lead," and that when there is no load on the machine the brushes will occupy a normal position. It is necessary to have a starting resistance in series with the armatures of all motors, whether shunt or series, because, if not, the armature would produce a dead short-circuit. Some motors have their starting switch so arranged that when the machine has got up a fair speed the starting resistance is · shifted from the armature coils into the shunt or field coils so as to strengthen the armature current and weaken the field.

In a series-wound motor on a constant-pressure circuit, the pull on the armature bars is in proportion to the amount of current flowing, and since the current is greatest at the starting moment, hence the maximum pull is exerted by the motor at that moment, because there is practically no speed or motion of the motor, and so there can be no back pressure generated. The field of the series motor is not constant like that of a shunt-wound motor. In a shunt motor, the current passing round the field coils is constant, and so at the moment of starting there is a heavy current only in the armature, the strength of the field remaining constant whether at the start or at full speed ; but in a series motor, the same current passes through the field coils as passes through the armature. Consequently, when a series motor is started, there is a heavy rush of current for a few seconds through both field and armature coils; hence the statical pull is enormously increased. In addition, a series motor does not run at a constant speed. Owing to these features, it is eminently suitable for traction work, and is almost universally employed for that purpose. Like the shunt motor, it is necessary to insert a starting resistance in the circuit of the motor; none is required in the field coils, because they are in series with the armature. So the starting resistance, the field coils, and the armature are all put in series, making one circuit; then when the motor has got up sufficient speed to generate a back pressure, the starting resistance can be taken out, if requisite. A caution might here be given that the load of a series motor should never be thrown off when the current is on or the motor in circuit, otherwise it will act like an engine without governors: it will increase its speed to a dangerous extent.

Compound-wound motors are not of much use, and are not so safe to run as a plain shunt-wound motor; when they get overloaded the action of the series coil interferes with that of the shunt.

In dealing with motors, very careful distinction must be made between the *maximum efficiency* of a motor and its *maximum work*. It was explained that the faster a motor runs the nearer does the back pressure generated approach to the value of the supply pressure; and since the current flowing is proportional to the difference of these two pressures, it is clear that when the motor is running at its maximum speed there is a minimum of current flowing, and the power absorbed is also a minimum. If it were possible for the motor to run so fast that it would generate a back pressure equal to the supply pressure, then no current would flow, and therefore no power would be absorbed, but the fact of running proves that some small amount of power *must* be absorbed in order to overcome the frictional resistance of the machine; hence the limit of the speed is such that the back pressure is slightly below the supply pressure. The motor is then doing no external work—only running idle—and the efficiency is then at a maximum—perhaps 98 per cent.

Taking the other extreme view of the case, let the motor be so overloaded that it cannot move; there will be no back pressure, but a heavy current will flow, limited to a safe amount by resistances. All the power absorbed is converted into heat in the coils of the machine, and again there is no external work done by the motor, because it is stationary. The efficiency in this case is at a minimum; it receives a large amount of power, absorbed in heating the coils, and gives none out. Hence, when standing still and running at its maximum speed, no useful external work is done. Therefore all external work must be done by the motor at speeds between these two conditions. By experiment it is found that the maximum amount of external work is done by the motor, when the motor runs at such a speed that the current flowing when the motor is stationary is reduced to one-half of its value—that is to say, the maximum work is done when the motor runs at a speed to develop a back pressure equal to one-half that of the supply pressure, so that signifies that when a motor is doing its greatest amount of work, it is working at an efficiency of only 50 per cent., and that the efficiency of a motor varies inversely to the amount of work it does, within certain limits.

It is useful to have a few of the most characteristic features of motors collected together, so that the different conditions of working that there are between a shunt and series wound motor may be easily perceived. The following refer to motors that are run from off constant-pressure mains.

The Shunt-Wound Motor (1) can be used only on a constant-pressure circuit; (2) the current in the field coils remains constant whatever the speed or the load may be; (3) the current in the armature coils decreases as the speed increases, or as the load decreases; (4) the torque or pull of the armature is proportional directly to the current in the armature; (5) the speed remains nearly constant with varying loads;.

(6) not suitable for starting with dead load on, or for mounting a grade.

The Series-Wound Motor (1) can be used on either constant-pressure or constant-current circuits, but chiefly is used on the former ; (2) the current in the field coils decreases when the speed increases, or when the load decreases ; (3) the current in the armature coils decreases when the speed increases, or when the load decreases ; (4) since the same current flows through both field and armature coils, therefore the torque of the armature will vary (within certain limits) with the *square* of the current flowing ; (5) the speed will vary inversely as the load ; (6) suitable for starting with dead load, or in mounting grades.

The E.M.F. that is developed by either a dynamo or a motor is proportional to the speed of the machine, and the strength of the current in the armature coils measures the amount of pull to which the armature is subjected. The product of pressure into current gives electrical power, and, similarly, the product of speed and pull gives mechanical power, so that generally, and in particular with reference to series-wound machines, pressure may be said to correspond with speed and current with mechanical pull. In the case of a dynamo, it is the armature current that strives to drag round the magnetic field in which it revolves ; and in the case of a motor, it is the magnetic field that exerts a pull on the armature current and so causes the armature to revolve. The torque of a revolving shaft is obtained by multiplying the pull by the radius of the shaft, and the power transmitted by the shaft is obtained by multiplying the pull in pounds by the peripheral speed in feet per minute. Thus, if P be the force or pull acting on the shaft, and r be its radius in feet, and $n =$ the number of revolutions per minute, then $P \times 2 \pi r n =$ foot pounds of work done per minute, and dividing by 33,000 gives us the horse-power.

Starting and Stopping Motors.—In starting electromotors, a little more caution is required than in starting dynamos, if such a statement can be made ; the reason for this is that the armature coils may get badly burnt if the requisite starting resistance is not operated in the right way. If by accident, however, the supply current is switched right on to the armature on starting the machine, it will be found that the safety fuse will blow before any serious damage will be done ; but that is not the only thing that will

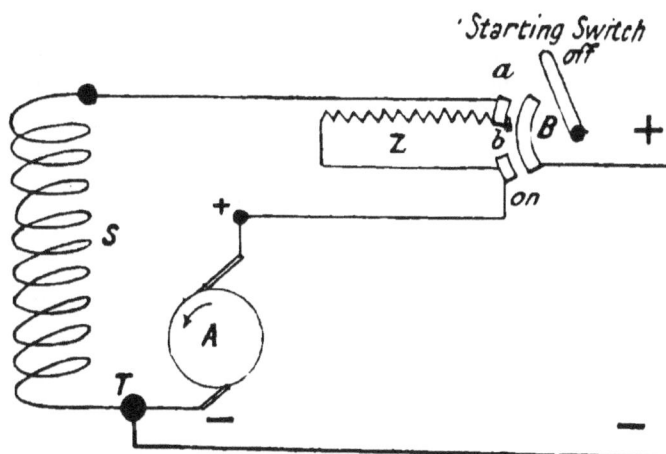

FIG. 29.

happen. A short-circuit of this kind will always burn or damage the commutator and brushes to a greater or lesser extent, according to whether the fuse is quick-acting or not.

Fig. 29 illustrates a common way of connecting up a shunt-wound motor on a constant-pressure circuit, the motor being belted to a ventilating fan, which it drives at a constant speed. The positive cable of the supply circuit is led to the starting resistance and connected up to the terminal, B, of the starting switch ; Z indicates the coils of the starting resistance, which

R

usually consists of a coil of iron wire wound closely on an insulated cylindrical core. On the other side of the switch there are two studs, indicated by a and b; one end of the resistance coil is connected on to a, and the other end is connected on to b. The stud b is marked " on," and from this stud a connecting wire is taken to the positive brush of the motor; one end of the shunt coils, S, is joined to a, and the other is joined to the binding post, T, of the motor. Two other wires are also attached to T—one from the negative brush of the motor, while the other is the negative cable of the supply circuit. The connections are now all complete, and we can easily trace the course of the current. Entering at B, it divides, part going through the shunt coils and part through the armature; the two circuits unite again at T, and by means of the negative cable complete the circuit. When the motor is not running, the switch must be open, and touching neither a nor b; so that before starting the motor, it is advisable to start it into motion by hand. This is not really necessary, but it simply prevents any undue strain on the motor. Having oiled up the machine and set down the brushes with a slight *backward* lead, the switch is first put on contact a; the result of this is that the resistance, Z, is put in series with the armature. Consequently a small current only flows through the armature, whilst the full pressure of the circuit is put on to the shunt coils, so that there is at the start a strong magnetic field. The greatest precaution must be taken to place the switch on this starting peg, as it is called, because if the switch is pulled over straight on to the " on " position without being first put on the starting pegs, the result will be that the armature will be burnt, unless the fuse protects it, since, when standing still, owing to the low resistance of the armature, it forms a dead short-circuit; but by using the

·starting peg only a small current is allowed to flow through the armature. The motor will then gradually get up speed, and at the end of half a minute or so the speed of the motor will generate a counter E.M.F.; then the switch can be pulled over to the "on" position, or stud *b*, as the counter E.M.F. of the motor will prevent too great a current flowing through the armature, so that when the switch is at "on," or on *b*, the starting resistance is taken out of the armature circuit and put into the shunt coil circuit, since not such a strong magnetic field is required as was required when starting the motor. To stop the motor, simply pull the switch right off sharp. There are other ways of connecting up shunt motors, but the above is one of the best and simplest where the load is constant.

Caution.—If there is a main switch in addition to the starting switch, and it is necessary for some reason to stop the motor by the main switch, the starting switch must be broken *before* putting in the main switch again, otherwise there will be a dead short on the armature.

For running a series-wound motor off constant-pressure mains, the motor is started by inserting the whole of the starting resistance into the motor circuit ; then, as it gets up speed, the resistance is gradually taken out by moving the switch forward. The more the resistance that is taken out, the faster will the motor run. To stop the motor, .gradually insert the resistance again.

CHAPTER V.

Illuminating Power of Arcs—Consumption of Carbon—
Diffusion of Light—Fixing and Trimming—Arcs in
Parallel—Town-Lighting—Incandescent   Lighting—Life
and Efficiency.

Illuminating Power of Arcs.

It was stated previously that we have every reason to
believe that there can be no light without the presence of
heat. Heat is a mode of motion and a source of energy, so
is light, but the amount of heat energy that we are capable
of converting into radiant light energy is only an extremely
small fraction of the total. Study Nature's animate display
of the conversion of energy into light, a living machine—the
glow-worm. Here, by some secret laws unknown to us, we
have the most wonderful example, one that dwarfs our most
ambitious productions into nothingness. The enormous loss
incurred by producing electric energy from the heat energy
given off by burning coal has already been mentioned. The
very best machinery and apparatus will only yield us 10 per
cent., and great is the problem to be solved that will enable
us to produce electric energy direct from coal, instead of
being compelled to employ those terribly wasteful converters—
the boiler and steam-engine.

But a far more important problem to tackle is how to
convert electric or heat energy direct into radiant light
energy without the enormous and almost, we may say, total
loss our present methods give ; for, whereas, in the first case,.

we can utilise one-tenth of the heat energy contained in the
coal, in the second case, so far as we can judge, we do not
obtain anything like this, but only a minute portion of the
whole.

After this digression we will return to practical matters.
There are two chief ways of converting electrical energy into
heat energy and then into light — first, by producing an
"electric arc," second, by raising a refractory substance to
incandescence, the first being known as the " arc lamp," and
the second as the " incandescent lamp." In both cases
carbon is used, which is raised to such a high temperature by
the passage of the electric current through it, that it becomes
white hot, and so emits lights. We will deal with the arc
lamp first, and then with the incandescent.

The principle of the arc lamp was explained in the
opening paragraph, and it is only necessary to add that the
mechanism of all arc lamps, simple or complicated, is for the
purpose of maintaining the two carbon rods at a definite and
fixed distance apart, by feeding one or both forward, as they
slowly consume away. In modern lamps the mechanism is
controlled by magnets and solenoids, energised by the same
current that flows through the lamp. Scores of arc lamps
have been invented and put upon the market since the first
display of the arc light, but very few of them were of any
good, and well has the law of the " survival of the fittest "
been exemplified. The domain of arc lighting is perfectly
distinct from that of incandescent lighting, the arc lamp
being essentially one for outdoor illumination, where for
large areas it stands without a rival, in particular such places
as railway sidings, goods yards, streets, smelting and foundry
works, canals, docks, quay sides, bridges, naval and military
purposes, etc.

It is now well understood that the light emitted by the arc
lamp is almost entirely derived from the incandescent and

glowing surface of the crater formed at the bottom of the positive or upper carbon. This "crater," as it is called, is a hollowing out of the carbon rod, and is somewhat conical in shape when it is well formed, and it seems to be the locality where the work is done, where the heat energy due to the electric current is dissipated and converted partly into radiant light energy. The work done consists of raising the carbon end to such a degree of temperature that it volatilises ; the carbon vapour so produced streams across the air gap between the positive and negative carbon rods, and settles upon the tip of the latter carbon. The glowing state of the tip of this lower or negative carbon is believed to be due probably to the settling of the white-hot carbon vapour, and also to its proximity to the intense heat of the crater above. Certainly a small amount of light is emitted from the tip of the negative carbon, and also from the flame of the arc that plays between the two carbon ends.

There are two common sizes of arc lamps, called by the trade a 2,000 nominal candle-power arc, and a 1,200 nominal candle-power arc. It must not be imagined that the first gives 2,000 c.p. and the second 1,200 c.p.— far from it ; very often the candle-power is anything but the nominal. Besides these two sizes, arc lamps are made in a number of other sizes, one or two being smaller, a sort of miniature arc lamp, while the sizes above go up to very great candle-power ; these, however, are used as search-lights, or projectors, as they are sometimes called. The remarks following will be confined solely to the two above-named sizes. The 2,000 nominal candle-power arc lamp, named a "full arc," takes usually about 10 amperes in current, at a pressure of 50 volts—that is to say, 500 watts, or half a unit. So that for every three-quarters of a brake horse-power given off by an engine one "full" lamp could be driven. The maximum actual candle-power yielded by this.

sized lamp is difficult to name ; in fact, nobody seems to know exactly, photometric tests applied to the arc light being very elastic. Between 800 c.p. and 1,000 c.p. is considered to be near the mark, while the average candle-power, or the average taken for all angles of observation below the lamp, is probably not much more than 500. The 1,200 nominal candle-power arc lamp, named a "half arc," takes from six to seven amperes in current, at a pressure of 50 volts, or, say, 330 watts, or one-third of a unit, so that for every half brake horse-power given off the engine one "half" arc could be driven. The maximum actual candle-power may be taken somewhere about 500 to 600, and the average candle-power does not seem to be much above 300.

When arc lamps are used for indoor illumination they should only be fixed in large and lofty structures, so that the powerful light can be dispersed over a floor area unbroken by any obstructive features, such as partitions, stands, etc. It is a common practice, particularly in London, to cram half-a-dozen or so arc lamps into a small shop. The effect is brilliant certainly, but rather startling ; probably the blaze of light produced acts on customers as the candle on the proverbial moth. Sufficient light is one thing, but an unbearable glare is something altogether different ; these displays, however, have one redeeming character : they serve to light up the streets close by very effectively. The following tabulation gives an idea of the illuminating powers of arcs at different heights, four degrees of intensity being given ; the arc in all cases being naked—i.e., having no globe over it. The first row, A, denotes an illumination just sufficient to see the way about, and only gives light enough to work by, when close by the lamp. This degree would serve for lighting up a railway track, canal banks, etc., where the lamps are fixed a good distance apart ; it is too expensive to light up well miles of track, so that the above suffices for its object.

The second row, B, denotes a degree of illumination so that ordinary work can be carried on, and is applicable to railway sidings, sheds, docks, quaysides, foundries, streets, etc. The third row, C, is a good strong light suitable for all ordinary indoor lighting, whether railway stations, halls, mills, shops, etc. The fourth row, D, denotes a stronger light, for places where an extra good light is wanted.

TABULATION 24.

Intensity of light.	Half arcs.			Full arcs.		
	Height in feet.	Radius in feet.	Area in square feet.	Height in feet.	Radius in feet.	Area in square feet.
A	30	114	80,000	40	170	100,000
B	30	68	16,000	40	76	20,000
C	15	24	2,000	15	34	4,000
D	15	17	1,000	15	24	2,000

The intensity of the light given off from the arc depends upon the point of observation. Imagine a circle described vertically round an arc lamp, having the arc as a centre ; then the light given off depends upon the angle that the point of observation makes with the vertical line drawn through the lamp or centre of the circle. At an angle of 90deg., or at right angles—that is, looking at the lamp in a horizontal line with the arc—the light is about a quarter of the maximum. Above the horizontal line the light gets feebler very rapidly, because the crater then becomes hidden from view. At a vertical point above the arc there is no light whatever. Taking a downward course, we find that the light gradually increases in intensity, and when the angle is 60deg. from the vertical or 30deg. below the horizontal line the increase becomes very marked, the intensity being about three-quarters of the maximum ; continuing lower down, when 45deg. is reached, or midway between the horizontal and vertical lines, the maximum intensity is yielded. Roughly speaking, the best part of the light lies between 30deg. and 60deg. below the horizontal, thus making 30deg. of light described as a belt round the arc ; the light rapidly losing its intensity, whether passing beyond or within this belt. Beyond 60deg. below the horizontal the bottom carbon intercepts more and more of the light rays; at 90deg. below the horizontal, or directly underneath the arc, there is no light—just as there was no light directly over the arc.

The direction of path of a ray of light emanating from a source is defined by its "angle of incidence." The angle of incidence signifies the angle that the ray in question makes with a vertical line dropping down from the centre of the source of light. The top diagram in Fig. 30 illustrates rays having angles of incidence of 30deg., 45deg., 60deg., 70deg., and 80deg., the horizontal line being, of course, a right angle, or 90deg. They also show at what points they touch the ground-line for various heights of the lamp. A, B, and C denote three ground-lines, thus giving the arc three different heights, in the ratio of 1, 2, 3. The innermost ray of the belt of maximum light previously mentioned has evidently an angle of incidence of 30deg., because 60deg. below the horizontal is the same as 30deg. out of the vertical. Similarly, the outermost ray has an angle of incidence of 60deg. These two angles of incidence are drawn in thick lines to mark the borders of the belt, and the base which lies between the borders is shaded for the three ground-lines, A, B, C. It will be noticed that the ray having an angle of 80deg. touches the ground-line some distance from the lamp; in addition to this, the light given at that angle is much below the maximum, so that the illuminating power is only feeble : therefore it is not worth while receiving rays that have an angle of incidence greater than 80deg.

An arc lamp forms an elevated centre of the illuminated circle, and it is interesting to examine into the relations between heights and distances, etc. We will assume that the arc is at such a height that the farthest point to be illuminated is reached by a ray having an angle of incidence of 60deg.

Let the height of the arc above the ground be, say, 10ft., as shown by the line, $b\ c$. Then, since the farthest point illuminated is where the ray of 60deg. reaches the ground, hence the horizontal distance, $c\ P$, will represent the radius of the illuminated circle. This distance, $c\ P$, we will call the "radius distance," to distinguish it from the angular distance, $b\ P$. Knowing the angle and the height, it is easy to find the radius distance, which is obtained from the formula :

Radius distance = height × tan A ;

also,   Height = radius distance × cotan A,

where tan A and cotan A represent the tangent and cotangent of the angle of incidence of the ray. In the present case the angle of incidence is 60deg., and the values of tan 60deg. and cotan 60deg. are respectively 1·73 and ·578. Inserting these values, we have

Radius distance = height × 1·73,

and   Height = radius distance × ·578.

So if the height, $b\ c$, be 10ft., then the radius distance, $c\ P$, will be 17·3. The length of the path of the ray can also easily be found when the height and the angle is known, since it equals the height divided by the cosine of the angle of incidence. Now cos 60deg. is ·5 ; hence $10 \div ·5 = 20$, so that the path of the ray denoted by $b\ P = 20$. It may be put down as a rough rule that an arc will illuminate well for a radius distance equal to double its height, because no rays beyond 64deg. are then required.

Rays having an angle of 80deg. will illuminate at a radius distance equal to about five and three-quarters the height of the arc from the ground. For a ray at a fixed angle, its radius distance and its length of path will be directly proportional to the height of the lamp, but if the candle-power

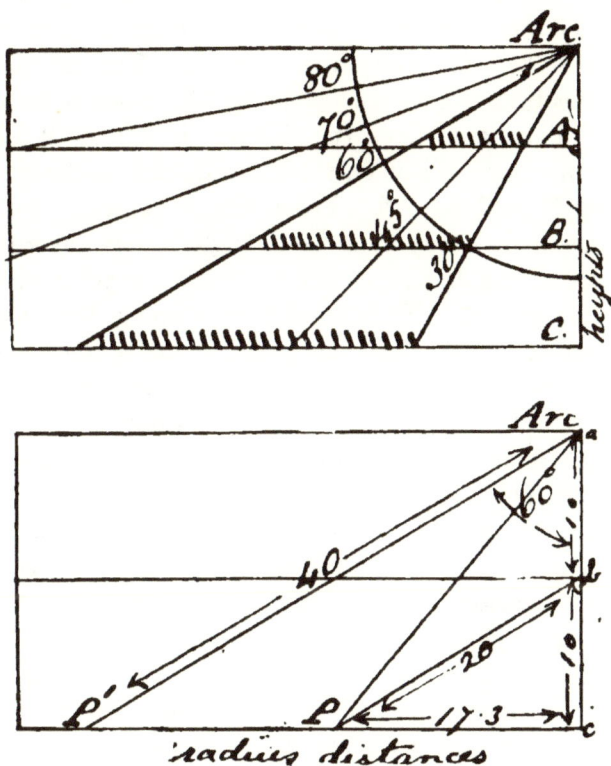

FIG. 30.

remains the same, the illuminating power of the ray where it reaches the ground will diminish very considerably as the height of the lamp increases. In the lower diagram of Fig. 30, when the height, b c, is 10ft., the radius distance, c P, is 17·3ft. for a ray of 60deg. as shown, and the length

of the path of the ray b P is 20ft. Suppose at the point
P, where the ray falls on the ground, that the illuminating
power or intensity of the ray is equal to 10 c.p., allowing
that the light given off the arc at this angle is three-quarters
of the maximum; now double the height of the lamp,
making it 20ft., the length of the path of the ray will now
be doubled and the new point, P', where it reaches the
ground will be 40ft. away from the arc. The intensity of
light varies inversely as the square of the distance from the
source of light, and since the distance is doubled it follows
that the intensity will be only one-fourth, and so the candle-
power at P' will be 2·5. From this it is seen that the
candle-power of a ray at a fixed angle varies inversely as
the square of the height of the lamp at the point where it
touches the ground. Now, instead of keeping the angle of
the ray constant, keep the same point, P, under consideration
when the height is varied; then, when the height of the
lamp is doubled, the angle of the ray falling on the same
point, P, will be much smaller. To find how the intensity
alters at the point P, we must first obtain the lengths of the
two paths, P b and P a.   P $b = 20$, and P $a = a\,c \div \cos$ P $a\,c$.
We have now to find out what the new angle P $a\,c$ is. To
do this, we already know that the radius distance $\div$ height
$=$ tan of the angle of incidence, therefore P $c \div a\,c =$ tan P $a\,c$;
that is, $17\cdot3 \div 20 = \cdot865 =$ tan 41deg., nearly. Having now
obtained our new angle, all that is wanted is the length
of the path P a, and, as given above, P $a = a\,c \div \cos$ 41deg.
$= 20 \div \cdot754 = 26\cdot5$; so that the path P b, representing the
old ray of 60deg., is 20ft. long, while the path, P a, of the
new ray of 41deg. is 26·5ft. long. The intensity varies
inversely as the square of the distance, hence the candle-
power at P, when the height of the lamp is 20ft., will be
$10 \times 20^2 \div (26\cdot5)^2 = 4,000 \div 702\cdot25 = 5\cdot7$ about.

Finally, we must make a correction, because the light

given by a ray of 41deg. is of greater intensity than what is given by a ray of 60deg., other things being equal. At 60deg. the light was assumed to be three-quarters of the maximum, and at 41deg. the light may be taken to be almost at the maximum, so $5\cdot7 \times 4 \div 3 = 22\cdot8 \div 3 = 7\cdot6$ = candle-power at the point P, when the arc is at a, or 20ft. high.

Consumption of Carbon.

It will be found in practice that the positive or upper carbon burns away at twice the rate at which the negative or lower carbon does. The proportions are not exactly 2 : 1 ; in different lamps and with different kinds of carbons the proportion varies slightly, but, as a rule, they will all come pretty close to the above ratio, and when trimming, the lamp trimmer always allows double the length of positive to the length of negative. The writer obtained the following rate of consumption of carbon, being the average of a number of tests made on a well-known " full " arc lamp :

Positive carbon, $\frac{5}{6}$, or ·833in. per hour.

Negative   ,, $\frac{3}{8}$, or ·375in.   ,,

Current being 10 amperes.

Pressure being 40 volts.

Energy consumed by lamp being 400 watts.

Diameter of carbons being 13mm.

With a full arc lamp, or one taking 10 amperes, the best size of carbons to use is 13mm., or $\frac{1}{2}$in., in diameter. With a half arc, or one taking six to seven amperes, carbons of 11mm., or $\frac{7}{16}$in., in diameter, should be used ; for a small lamp taking only five amperes, say, about 9mm. is suitable, whilst for a very powerful lamp, such as a search-light of 25,000 c.p., and taking a current of probably 50 amperes, the size would be about 24mm., or close on 1in., in diameter

The rate of consumption of these larger-sized carbons is less than the smaller-sized ones, the burning being slower as the carbons increase in thickness. Carbons smaller than 9mm. for five amperes, and those larger than 40mm. for, say, 130 amperes, do not burn well, it being very difficult to obtain a well-formed crater, consequently the light is unsteady. It is the custom now in the best kind made to insert a central soft core in the positive carbon ; these are called cored carbons. They are naturally more expensive than the other solid kind, but they give a better, softer, and steadier light. The soft core tends to keep the crater in the centre of the carbon, and promotes a more regular burning ; the light also is purer and whiter. There is a great art in manufacturing electric light carbons, each maker having some little secret of his own in compounding the ingredients that form the paste. Gas-retort carbon, or gas coke, as it is called, is crushed into a fine powder and then mixed with several things, as pitch, oils, etc. ; the compound is passed through a number of processes, then subjected to great pressure, and finally baked in the furnace. Provided a good lamp be employed, the character of the carbon soon proves itself ; a bad and impure specimen burns irregularly, splutters, and is continually breaking its arc, and so causes the light to fluctuate up and down, and, above all, the light given off is often a greenish-red or a purple colour, making those on whom it falls have a very bilious look. A good carbon, however, acts very differently : it burns steadily, with little waste, the light is pure, full and white, and perfectly steady.

Coated or coppered carbons are those which are coated over with a thin deposit of copper, to conduct the current better. A single-carbon lamp means one that has only one pair of carbons, and a double-carbon lamp one that has two pair, so that when the first pair is burnt out, the lamp changes over on to the second pair. The usual length that carbons are cut is about 16in. for positive, and 8in. for negative, so that the single lamp will burn for 16 hours and the double lamp for 32 hours. The consumption will not reach 1in. per hour for the positive, as the carbon stumps will testify; in any case the carbons should never be allowed to burn down to less than ¾in. away from the holders, otherwise the holders will get heated and burnt. The depth of holders is about ½in., so the stumps should not be less than 1¼in. long when taken out.

Short and Long Arcs.—There are two distinctive kinds of arcs known in arc light illumination, the "short" arc and the "long" arc. The "short" arc, as its name implies, occurs when the two carbon pencils are only slightly separated, say ·75mm., or ·03in. Evidently, with such a short distance, only a low pressure is required, say 25 to 30 volts, but at the same time, to have the proper heating effect, what is decreased in pressure must be increased in current in order to obtain the requisite number of watts, so that, while the pressure is only one-half the pressure used in "long" arcs, the current is doubled. There are several serious objections to the "short" arc—they do not burn steadily, they are noisy, and in a system of series lighting the line loss of energy is very heavy, since doubling the current signifies four times loss of power. The "long" arc is about 2mm., or ·075in. long: this is two to three times the length of a "short" arc. This form of arc usually burns very steady and silently, and absorbs from 40 to 45 volts, with only half the current that a "short" requires.

The action of the carbons when burning indicates whether the lamp is burning properly, and to some degree tells us what is wrong. For example, if the arc emits a hissing sound it may be put down that the feed is defective, since the arc is too short and the feed of the lamp is too rapid.

On the other hand, when the arc is too long, or the current too strong, the arc will flame. Spluttering of the arc may arise from several causes, chief among which is that of impure carbons. Nearly all arc lamps are now made to burn with "long" arcs, as they are more satisfactory.

Diffusion of Light.

A number of people complain that the electric light, whether in the form of an arc lamp or incandescent lamp, is irritating and painful to the eye when looking at it. When the lamps are entirely uncovered, and the eyes are allowed to rest on the naked arc or filament, as the case may be, the above troublesome effect is not to be wondered at. It must be remembered that the light is of very great intensity, irrespective of the quantity of light, and that being so, concentrated rays require dispersing or diffusing. Daylight is light in the most perfect diffused state, and the source, the sun, need not be visible to our eyes at all. Moonlight, which is borrowed sunlight, is a beautifully softened light, although the shadows are strong. Contrast these with our methods of artificial lighting. We produce light by sprinkling about a greater or smaller number of points or concentrated sources of light, the light being intense close by, and diminishing rapidly at a distance. If we were to break up these points of light into a countless number of infinitely small sources, placed close together, we should then obtain a diffused effect. An ideal illumination of a room would be to have the whole of the ceiling to throw off a subdued and diffused light, thus converting the irritating and unpleasant points of light into large luminous surfaces.

A most simple example is to compare an incandescent lamp and a gas jet. It is much easier to look at the latter than at the former. Why? Because the electric lamp gives its light from off a slender thread of carbon raised by intense heat to a white-hot state. The luminous surface is here extremely small, but extremely intense. On the other hand, the light from a gas jet is given off from a broad surface, and the light is not intense. The quantity of light in each case is about the same : the former being a concentrated light, while the latter is less concentrated, and consequently more diffused.

We have not made one jot further progress in the art of illumination since the days of the ancients, when perfumed oils were burnt in their vessels ; and so far as gas is concerned, we have taken a retrograde step, for whereas the oil vessels were costly articles and the burning of good oils produces a pleasant and soft light, the ugliness of gas brackets and the foulness of the gas burnt are universally known.

The intensity of the light of the electric arc is probably greater than any other artificial source of light. After gazing at a naked arc for a few seconds the eyes will be almost blinded, just as if one had looked at the sun. The temperature of the crater has been variably estimated, but 3,000deg. to 6,000deg. C. is near the right number. The light rays that are emitted from the glowing crater are only about 10 per cent. of the total rays, the other 90 per cent. being invisible, or dark heat rays, and yet this source of light gives many more visible light rays than other forms of artificial illumination ; for example, a gas flame gives only from 3 to 4 per cent.

Arc lamps are now rarely run without having some kind of globe over them, except when used for lighting up large open spaces belonging more or less to private concerns—like dockyards, works, goods yards, etc., where appearance is not considered provided they get the light. There are several different kinds of globes used on arc lamps, each of which absorbs so much of the light : all of them help to diffuse

S

the light and to lessen the shadows. By using very thick porcelain globes a great amount of light is absorbed or cut off, but against this disadvantage must be put the fact that the light available is well diffused and softened down, thus reducing the intensity and preventing deep shadows. The following figures give the probable amount of light that is cut off by different globes :

TABULATION 25.

Clear glass	10 to 20 per cent. loss	
Thin ground glass	20 to 30 ,,	,,
Thick ground glass	30 to 40 ,,	,,
Thin opal glass	40 to 50 ,,	,,
Thick opal glass	50 to 60 ,,	,,

So that the light can be toned down to any degree by using a suitable globe.

The City of London Electric Light Company use a lantern having panes of ribbed prismatic glass, whereby the arc light is greatly magnified and softened down, so that the light rays are well diffused.

The value of a light for any particular purpose depends upon its composite rays, and the superiority of the arc light for illuminating purposes can easily be proved to be much beyond oil, gas, or any other artificial source. We will, therefore, analyse different kinds of light, and briefly sum up their composition of rays, taking the sun as the standard.

The light from the sun is composed of a number of homogeneous colours mixed together in a certain proportion, and forming what is called a "spectrum." The light given by the spectrum of the sun is colourless, and is called white light.

The colour of any substance or matter is regulated by the kind of rays which it *absorbs* from the light by which it is illuminated ; for instance, a white substance, like snow or anything with a white surface, does not *absorb* any of the rays of the sun, but *reflects* them all back again, hence the snow appears colourless or white in daylight, because the light of the sun is white. A black surface, on the contrary, like coal, absorbs all the rays of sunlight, and so appears colourless or black. So when all the sun's rays are visible we have no colour, or a white appearance, and when none of the sun's rays are visible we have also no colour, or a black appearance. A substance having a yellow appearance signifies that all the sun's rays are absorbed except the yellow rays, and these are reflected back, thus giving the yellow tint. After this preliminary explanation respecting colours, we can now realise that a substance may present a different or modified colour when it is looked at in the light given from some artificial source.

The light given by gas, oil, or candles contains all the colours that the sun contains, but they are mixed in a different proportion; this is the same with the light of the electric arc. The tabulation below shows how the various coloured rays mentioned are distributed in the three illuminants—the sun, gaslight, and the electric arc.

The electric arc light is much nearer the light of the sun than any other artificial light; consequently, the natural colours of things are seen better by the electric light than by gaslight. Gaslight is particularly rich in the *red* or *heat* rays, while the electric arc is particularly rich in the *violet* or *chemical* or *actinic* rays.

The arc light is invaluable when colours require matching, because, with the exception of a very few shades, all the colours and shades known to us can be seen just as they would be in the light of the sun or daylight. A certain eminent silk firm made a thorough trial of the value of the electric arc in their mills, and they found that out of about 650 different colours and shades, there were only seven that could not be matched in the light of the electric arc. The

s 2

writer cannot recollect what they were, but bronze was one of them.

TABULATION 26.—Colours of Rays.

Degree.	Yellow.	Red.	Violet.	Blue.	Green.
Rich	Gas.	Gas.	Arc.	Arc.	Gas.
Poor	Arc.	Arc.	Sun.	Sun.	Arc.
Very poor.	Sun.	Sun.	Gas.	Gas.	Sun.

Everybody knows how impossible it is to distinguish between shades of blue and green in gaslight. The above tabulation explains the reason why ; there gas is shown to be very poor in blue rays, but very rich in green rays. A green substance reflects a small number of blue rays in addition to the green rays, and a blue substance reflects both blue and green rays. Therefore when a blue cloth is inspected by gaslight it will appear more or less greenish, because gaslight has plenty of green rays, but very few blue. On account of the great number of violet rays present in the arc light, the arc lamp is of great use in photographic work, such as the printing of negatives, etc.

Fixing and Trimming.

The structural arrangement of a building has a great deal to do with the hanging of arc lamps. The positions in which the lamps are placed are mostly regulated by the position of the girders, columns, etc., so that they must not be hung anyhow and anywhere. When determining these positions, provision must be made for getting at the lamps easily and quickly for repairs, trimming, cleaning, etc. ; difficulty of access causes loss of time and loss of money, and, in addition, is often a perpetual source of annoyance to the occupiers of the place that is lighted. Buildings that are roofed by a lofty arch are particularly troublesome to deal with. Take the case of a railway station

·as an example, where the span is built of iron and glass. This must be wired without incurring the expense of using scaffolding, and calls for a fair amount of activity and nerve on the part of those engaged in the work. The lamps should in these cases be fixed on either side of the centre of the arch, that is if the lighting arrangements allow of this being done. Another matter of importance, and one that will give plenty of scope for carrying out ingenious plans, is to wire the place in such a way that the length of cable used is brought down as low as possible. A great amount of cable can be wasted on jobs of this sort, unless a good economical wiring scheme is thoroughly thought out. Great precautions must be taken that there is good insulation everywhere. When running cable along iron girders, etc., porcelain insulators must be freely used, so that at no point does the cable come in contact with the ironwork.

Arc lamps are best hung with steel rope—No. 8 will do ; and since nearly all of them have one cable connected to the metal framework of the lamp, therefore the steel rope must pass through an insulator. In place of the iron hook at the top of the framework, some lamps have a little porcelain pulley.

Every lamp should be provided with a cut-out switch by means of which the lamp can be cut out of circuit. Should it be necessary to handle the lamp in any way when the current is on the circuit, it is of the most vital importance that the trimmer should switch that particular lamp out of circuit before attempting to touch it ; it is only the work of a second, and personal safety is then ensured. Through neglect of this simple precaution several fatal accidents have taken place—all owing to gross carelessness on the part of the trimmer ; because this is a most rigid rule in all arc light central stations, and need be so, when it is remembered that the working pressure is often 3,000 volts or so. An arc lamp can be hung, "fixed" or "movable." In the former case, the trimmer has to carry trestles from lamp to lamp, which involves a good deal of labour and loss of time. This is speaking of indoor lamps. For outdoor lamps, such as those for street-lighting, the trimmer only needs a light ladder, because the lamps are fixed on standards or projected from walls.

By going to a little more expense, indoor lamps can be made movable by using a lowering gear with a counterweight. In this case the lamp can be drawn down to the floor for trimming, etc., and then shot up again ; the one objectionable feature of this method is the presence of the "slack cable." A 10-ampere cable, well insulated, is about as thick as one's little finger, and when hanging loosely has an unsightly appearance ; to avoid this, several devices have lately been brought out to automatically wind and unwind this slack as the lamp is moved up and down.

An arc lamp should never be suspended by means of its own cables. The weight of the lamp should in all cases be taken by a means of suspension perfectly independent of the cables, and when the lamp is at its proper position the cable should have no more tension in it than to give a fairly taut and trim appearance.

When lamps are hung from posts, out of doors, in exposed positions, it is necessary to fasten the bottom of the lamp to its supports, otherwise it will sway about in a high wind.

Putting a fresh pair of carbons into a lamp, to replace those which have burnt away, is called "trimming," and the man who attends to this duty, and looks after the lamps generally, is called a lamp trimmer. In trimming, great care must be taken that the two carbons are put in their holders so that they are perfectly in line. The carbon-holders are made adjustable in most lamps, so that they can be shifted about. When a carbon does not fall in line, it will often be found that simply turning the carbon round a little will make it right. When trimming a lamp, always put in the positive first, and then allow the feed-rod to run down slowly. If the carbon is fixed right, it will glide through the aperture of the negative carbon-holder, thus proving that this one is centrally fixed; it only remains now to fix the negative so that its point is opposite the point of the positive. The brass rods of the lamp should never be cleaned with emery-cloth, because the constant rubbing would produce wear and the adjustment of the lamp would be spoilt; only paste should be used; at times when the rods have become very dirty, a little crocus-paper might be used, but only sparingly. Lamps that are fixed in foundries, smelting works, and all such places of this sort, get smothered with grime and dust. The mechanism of such lamps should be well covered up. When lamps are fixed anywhere near to the sea, such as on the promenades of seaside places, if they are only used for the summer season, they should be taken down and stored away when the season is over. When in use they will require well looking after, because they are exposed to stormy weather, and the sea-brine in the atmosphere tends to rust the mechanism of the lamp. The cables, being also affected by sea-air, should always be buried underground if possible, or, at least, protected as much as possible from the atmosphere, and, in particular, they should be placed where no sea-water or sea-spray can wet them.

Arcs in Parallel.

In an incandescent system of lighting it is very often required to run a few arc lamps. The lamps are then run two in series across the mains—any number of pairs of lamps can be run like this. When the pressure is above what is required for two lamps in series, the surplus is absorbed by a small resistance coil placed in series with the lamps, so that the total electrical power consumed is the product of the circuit pressure into the current the lamps take; of this power, evidently that absorbed by the resistance coil is wasted. For example, suppose we have a number of incandescent lamps working at a pressure of 110 volts—two "full" arc lamps take about 10 amperes—and a pressure of, say, 42 volts each, there is then $110 - 84 = 26$ volts left, and since the same current passes through both lamps and resistance coils, therefore $26 \div 10 = 2.6 = $ the resistance, in ohms, that the resistance coils must have in order that the 26 volts shall be absorbed in forcing a current of 10 amperes through it, because $10 \times 2.6 = 26$. The above figures relate to some arc lamps run off an incandescent plant now at work. Lamps that are run in parallel always burn better when there is a good margin of pressure, even though this extra pressure is absorbed by a resistance. To run a single lamp on an incandescent circuit would require a large resistance coil unless the circuit had a very low pressure, such as 70 volts, so that for the usual pressure of incandescent circuits, 100 or 110 volts, more power would be wasted in the resistance coil than would run a second lamp.

In the event of one of the two lamps that are run off the mains going out, or otherwise failing to act, the second one goes out as well, there being no cut-out as a rule on lamps adapted for parallel running; the reason of this is that the remaining lamp would have double the pressure, and hence its current would be doubled, which would at least damage the regulating coils of the lamps, and in all probability burn them up.

Town-Lighting.

The greatest field for arc lighting is undoubtedly that of the illumination of the streets and public thorough-fares of towns, and now that some of our most important cities are having central stations put down, there is great prospect that the dull, miserable gas lights will quickly disappear, to make room for the brilliant light of the electric arc. Nobody can fail to mark the great contrast presented when passing from a district lighted by arcs into one lighted by gas—the latter in comparison with the former is a region of gloom. In England it is very seldom that one comes across a town that is decently lighted by gas, and even then it is confined to the few important streets. It is the effect of the contrast between gas and electric lighting that will do more to further the adoption of the latter than anything else. People that visit a town brilliantly lighted by the new illuminant will feel the gloom and semi-darkness of their own gas-lighted town in a tenfold degree, for when a higher standard is reached, no matter in what, the old one is dwarfed in value, and as the gas flame supplanted the rush-light in its lighting powers, so will the electric light supplant gaslight. Next to drainage, one of the very greatest blessing a town can have is good light, and plenty of it ; not merely for personal convenience in getting about in the streets, but for the whole welfare and improvement of the place. The town will score a decided advantage over its ill-lighted neighbours : business is brisker, the vehicular traffic is rendered safer and more expeditious, and the general tone of the place is raised. A brightly-lighted town tends to make the inhabitants more cheerful and smart. Ill-lighted towns drive away trade and visitors, throw a despondency over all, and nurse crime as darkness does noxious plants.

In the majority of English towns the central part consists of narrow, tortuous streets huddled together, forming the old nucleus of the town, and where the town is one that has opened out and improved, wider and straighter roads will radiate out from this centre. When going into the matter of lighting a town by means of arc lamps, the plan of the town

must be very carefully analysed and studied, and it is no use
trying to find a precedent: each town must be considered
and judged from its own advantages and disadvantages. No
two are alike : each has its own local peculiarities ; and
every street, every square, must be thoroughly considered
with regard to amount of traffic, public buildings, most
suitable distribution of light, and such like details, so that
the engineer must identify himself with the requirements of
the town as if he were an inhabitant, and so had to partici-
pate in the lighting arrangements he is about to introduce.
Planned-out roads, fairly straight, or with gentle curves, are
easy enough to light ; it is the rabbit-warren collection of
streets that cause trouble, for light will not go round a
corner, and since arc lamps are placed much farther apart
than gas lamps are, it becomes difficult to light up a street
that bends about. In these cases it is perhaps best to employ
arcs of small candle-power, so that they can be placed nearer
together than what larger sized lamps would be ; then in the
straight roads these large lamps, well apart, can be positioned
where their full light will be utilised. Small and side streets,
alleys, etc., are best lighted by incandescent lamps—16 c.p.
or 32 c.p., according to the amount of light deemed necessary;
lamps of 8 c.p. are of little use, being too feeble, and they
should never be used for street-lighting. Incandescent lamps
can easily be run off an arc lighting plant, and the combination of the two, judiciously placed, provide a flexible and·
satisfactory way of dealing with all the streets of a town,
from the main thoroughfares down to the insignificant side
street.

The positions of the gas lamps indicate best where the
light is mostly wanted ; they will also indicate by their
distance apart the relative degree of light required in various
parts of the town, so that although they do not illuminate
well, they provide a good guide as to how the light should

be distributed. The great art in fixing arc lamps in the streets is to always have one at a corner or junction of roads, and erected in such a position that, whether there be three or four or more roads meeting at that point, the light from the arc shall penetrate down all the roads or streets. It is at such corners and junctions that the use and value of the arc is shown, since the number of gas lamps which one arc will displace will increase with the number of roads which the arc commands. Suppose one arc will illuminate as far as three gas lamps at a certain distance apart, then one arc will displace three gas lamps along a straight road, and at three cross roads will displace five gas lamps, and at four cross roads as many as seven.

In small country towns very little traffic is in the streets after 10 p.m., and most of the inhabitants are indoors, and towards midnight scarcely a solitary person is about, so that at that time one-half of the arc lamps could easily be dispensed with and switched off. This would effect a considerable saving, for in addition to the decrease of the current and the saving of electric energy, the consumption of carbons in the lamps is reduced probably to one-half, and the lamps will require less trimming, thereby saving the trimmer's time. The incandescent lamps could not very well be diminished ; being small lights, they would all be required. In the United States, when a town enters into a contract with the electric company located there to light the place with arc lamps, the moon schedule is always brought in ; this means that whenever there is a full moon visible and an unclouded sky, the arcs can be switched off, since the moon supplies ample light : for every night on which the moon lights the town instead of arcs so much is taken off the contract price. The same might be done in England, but at much greater risk, because there is no knowing how the weather may change.

Where the roads are wide and the traffic permits it, the best position for the arc posts is in the centre of the road : the light is then distributed equally on each side. In the majority of places, however, it will be found that they must be placed along the kerbstone, like gas lamps, and fixed alternately on the right and left side of the road. When arc lighting was first introduced into the streets, wooden poles with a slight taper and painted dark green were used, the height being about 30ft. Now that the lighting is a proved success, it is better to erect more substantial ones in the form of light steel tubular poles or cast-iron ornamental standards. The arc lamp can be fixed on the top of the standard, or it can be hung from a bracket arrangement ; this latter way is useful when the lamp is at a street corner, since it can be projected over the roadway. Chelmsford, in Essex, was the second town in England to have its streets entirely lighted by electricity, arc lamps being used for the main thorough-fares, and 32-c.p. incandescent lamps for the rest. The main artery of the town is about a mile long, and takes a right-angled course; there are 20 arcs, and therefore 20 arcs to the mile. This gives an average distance apart of 88 yards, or 264ft. In the central portion of the route they are fixed 70 to 80 yards apart, further away the distances increase to 120 to 130 yards apart. The height of these arcs is 30ft. ; the writer does not know their actual candle-power.

Until lately there was no arc lamp that would burn satis-factorily with alternating currents, and there are now very few in the market that are of any use, so that there is an unexplored field open for the adaption of arc lamps for alternating currents. Since an alternating current flows first in one direction and then in another, therefore the carbon tips both become somewhat pointed, and no crater is formed in either, consequently as much light is thrown upwards as is thrown downwards, a very different case to arc lamps run with continuous currents. These upward rays are therefore wasted in street-lighting, so that small reflectors should be provided fixed over the arc, thus throwing the upward rays towards the ground.

There is very little reliable data concerning a comparison between an alternating-current arc and a direct-current arc. Some say the former gives more light than the latter for the same amount of power. There is certainly a great future before a good alternating-current arc lamp, particularly one that is adapted for running in parallel across the secondary of a high-tension alternating-current system, such as is mostly used for distributing electricity over wide areas. One curious thing about these lamps is that they emit a humming sound when burning. The cause of this is not known. It may be due to the alternate rapid heating and cooling of the carbons, promoted by the continual change of polarity. Cored carbons should be used for both top and bottom in these lamps.

Incandescent Lighting.

The light given off by the second kind of electric lamp—namely, the incandescent or glow lamp—is really due to the same cause as that which produces the arc light—*i.e.*, carbon raised to an incandescent or glowing state by means of electric energy. In the arc lamp the crater is white-hot carbon, and in the incandescent lamp the whole of the filament is made white hot. The reason why the light of an incandescent lamp is so much below that of an arc lamp is because the carbon filament of the former is only raised to about the temperature of melting platinum, say 1,800deg. C. or 2,000deg. C., whilst the crater of the arc is raised much above this, perhaps up to 6,000deg. C. ; certainly not below 3,000deg. C.

The incandescent lamp, as is generally known by this time, consists of a very thin thread of carbonised fibre, technically called the "filament," hermetically sealed within a pear-shaped glass bulb, which is exhausted of air as much as possible. The filament is bent parallel, somewhat after the manner of a hairpin, and sometimes has a curl at the bend. The two ends of the filament are jointed on to two platinum wires, which pass, embedded, through the neck of the bulb and make contact with two brass terminals. These terminals are fixed in plaster of Paris, thus forming a terminal collar to the lamp. Unlike the arc lamp, the incandescent lamp is seen to its best advantage for indoor illumination, and being of small candle-power the light can be distributed in any way desirable, and fixed in any place.

For the same amount of electrical energy spent, an arc lamp will give from eight to nine times the amount of light that an incandescent lamp will give. An arc lamp gives, say, 1,000 c.p. for 400 watts; this is at the rate of 2·5 candles per watt, or ·4 of a watt per candle-power. An incandescent lamp of 16 c.p. absorbs, say, 56 watts, and this is at the rate of ·28 candle per watt, or 3·5 watts per candle-power. These lamps are made with a great variety of candle-power, and also to burn at various pressures. The standard size is a lamp of 16 c.p. to run at 100 volts pressure, because this is the most common pressure adopted in consumers' houses. The same sized lamp can, however, be made to run at a higher or lower voltage, such as 50 volts or 120 volts. For indoor illumination, 8 c.p. and 32 c.p. are also greatly used, the smaller one being very useful for lighting places that do not require much light—such as corridors, staircases, small rooms, recesses, etc. ; the larger light is useful for places where an extra illumination is required—as shop windows, billiard tables, etc. The very large sizes—as 100, 200, and 500—are known as "Sun-beams," and are mostly used for lighting large rooms, entrance halls, or outdoor. These lamps are not used very much because their "life" is not long, and they blacken and decrease rapidly in candle-power. In addition to all the above, there are very small lamps of five, three, two, and half candle-power—3 c.p. is generally found in miners' electric hand lamps, where the lamp is connected up to an accumulator battery. The smallest lamps are used in theatrical effects, and in surgical and such like operations.

A lamp of same candle-power and efficiency consumes the same energy whatever voltage it may be made for, because the lower the pressure the greater must its current be. For example, a 16-c.p. lamp at 100 volts pressure has a filament of about 180 ohms resistance, and will therefore take a current of $100 \div 180 = \cdot55$ of an ampere. Suppose a similar lamp is required to run on a circuit which is at a pressure of 50 volts, evidently at this pressure only half the current would flow through the filament: therefore to have the same current the resistance of the filament must be halved, and this means doubling the sectional area. But although we have now the same current, we only use half the pressure, consequently only one-half the electrical energy, or number of watts, because $50 \times \cdot55$ is one-half of $100 \times \cdot55$. To get the same quantity of light we must utilise the same amount of energy. By again halving the resistance, or again doubling the sectional area of the filament, we get 1·1 amperes through, and then $50 \times 1\cdot1$ equals $100 \times \cdot55$; so that to run a lamp at half the voltage requires double the current, and the filament will have four times the sectional area, or double the diameter. Generally, then, as the working voltage decreases, the diameter of the filament increases. It is very easy to distinguish between an 8-c.p. and a 16-c.p. lamp when the candle-power number is illegible : the 8-c.p. lamp has a much finer filament than the 16-c.p. when both are made for the same voltage.

The following tabulation gives the current that various sized lamps take.

TABULATION 27.

Candle-power.	Volts.	Amperes.	Watts.
1	2·8	1·5	4·2
2	25	·32	8
5	50	·4	20
8	100	·32	32
16	100	·56	56
32	100	1·12	112
50	100	1·75	175
100	100	3·5	350
200	100	7·0	700
500	100	17·5	1,750

Of course there are a number of other pressures that lamps are made for besides 100 volts, but that is the most common one for incandescent lamps. Where storage batteries are employed in conjunction with machinery at private establishments, the circuit pressure is often fixed for 50, 60, or 70 volts. Then there are lamps made for 105 and 110 volts.

TABULATION 28.

Volts.	Candle-power.	Watts.
25	·4	14·0
29·5	·87	19·
34·8	2·47	26·9
40·0	5·1	35·9
48·0	12·6	46·3
49·0	15·0	50·5
50·0	15·8	52·7
52·5	20·5	57·5
52·6	28·4	64·5
59·5	39·3	72·9
62·0	50·7	79·9
68·2	74·5	96·7
72·5	103·2	107·5

A lamp should only be run at the voltage for what it is made. If this be not done, then when the voltage is lower the light will fall off enormously, and when the voltage is higher the light will be enormously increased, but the lamp will be quickly burnt out. To show how rapidly and out of proportion the light of a lamp varies when its voltage, and hence its current, is varied, the figures given in Tabulation 28 are the result of an experiment made on a 50-volt lamp of 16 c.p. taking 52 watts. By studying these figures it is seen that when the voltage has fallen two volts, or 4 per cent., the light has fallen no less than 21 per cent., or more than one-fifth. Approximately, the variation of candle-power is in proportion to the sixth power of the voltages—this is limited to within about 20 per cent. below and 20 per cent. above the normal candle-power.   Numerous empiric rules have been formulated for expressing the rise and fall of candle-power in terms of the voltage, but none of them are of much service for practical purposes, and the rough rule given above will be found sufficient for most purposes.   Upon applying it to the example just quoted, it will be found that it holds true and is almost exactly correct, for the ratio between the sixth powers of 48 and 50 is about 4 : 5, and four-fifths of 15·8 works out 12·6 ; similarly the ratio of the sixth powers of 50 and 52·5 is about 3 : 4, and 15·8 multiplied by 4 and divided by 3 gives 21, which is near enough to 20·5 given by the table.

An incandescent lamp of 16 c.p. will illuminate well a floor area of about 50 square feet when placed, say, 10ft. high, and will give a good ordinary light for every 80 square feet. A large percentage of the light goes downward when the lamp is fixed with its holder uppermost, and when the lamp is fixed on a table standard, or on brackets, so that the bulb is upwards, more light goes upwards than downwards. So it is essential to use reflecting shades if it is required to direct the light in any way and obtain its full benefit, such as for reading purposes, workshop benches, and all places where

T

each individual wants a small light to himself. It is in this case that the advantage and superiority of the incandescent is mostly proved over every other kind of light—even the arc lamp is then at a disadvantage. The lamp suspended by a flexible cable can be raised and lowered, as desired ; it can be placed within an inch or so of any work to be examined, and when provided with a hand support and length of flexible cable can be placed and carried about anywhere. For reading, perhaps the pleasantest kind of shade is green porcelain having a white enamelled interior surface. Rooms and halls having whitewashed ceilings require less illuminating power than those which have not. Similarly, rooms with light-coloured wall paper reflect more light than those with dark-coloured paper.

The illuminating power of all incandescent lamps deterio-rates with duration of burning, and the longer they burn, the more their light is diminished. When the lamps have reached a certain stage of dimness, it is uneconomical to keep them burning, and they should then be replaced by new lamps, the old ones being either broken or placed in some unimportant place where a decreased light is of little moment. A great deal of the deterioration of the light is certainly due to the blackening of the inside of the glass bulb, caused by the gradual disintegration of the carbon filament, which action deposits a fine film of carbon on the surface of the glass. There are one or two other causes, such as the thinning of the filament as it disintegrates. This increases its resistance, and so helps to diminish the light ; beyond this there is not much knowledge on the matter : it is suggested that the vacuum becomes less perfect. To retain the initial illumi-nating power of a lamp is the most important problem to solve in their manufacture, and to effect this the filament must be made durable, and able to withstand high tempera-tures. The glass bulb is made either clear or clouded : in the former case the filament is exposed to the eye, and is painful to look upon, owing to its intensity ; by having the glass clouded or frosted the filament is hidden, and the light is diffused over the bulb, hence giving a subdued and more pleasing effect. Any loss of light that may result from this kind is amply compensated by the increased comfort their use confers.

Life and Efficiency.

The average life of a 16-c.p. lamp is estimated to be close upon 1,000 hours. A few break in a few hours, but this is very seldom, and they are defective lamps : others have various lives, ranging from hundreds to over 1,000 hours, whilst others, again, have been known to last over 3,000 hours, but the majority of lamps work out to an average somewhere about 1,000 hours; hence this number may be taken when making calculations respecting their running cost. The following figures show how the candle-power falls off in different stages of their life, being the result of tests made at Cornell University on a 16-c.p. lamp worked at 100 volts pressure ; the decrease of energy consumed is also given :

TABULATION 29.

—	At start.	100 hrs.	200 hrs.	400 hrs.	800 hrs.
Candle-power	16	12·5	10·8	9·67	7·2
Watts	48·24	46·21	45·9	43·61	42·33
Watts per c.p..........	3·01	3·69	4·25	4·51	5·88

The efficiency of an incandescent lamp is often interpreted as the rate of initial electric power it consumes in comparison with the initial light given out ; thus, a lamp which takes only 40 watts, and gives 16 c.p., when new is termed a high-efficiency lamp, whereas a lamp that takes 64 watts, and gives 16 c.p., is termed a low-efficiency lamp. Normal

T 2

efficiency being generally granted when a lamp takes 56 watts, or 3·5 watts per candle-power. The term efficiency when employed in this way is very misleading, because, although a so-called high-efficiency lamp only consumes small power at the commencement of its life, it unfortunately has a short life, and the candle-power diminishes very rapidly; and the smaller that this initial power is, the shorter becomes the life of the lamp, and the more does its light diminish. On the other hand, by using a so-called low-efficiency lamp, the large consumption of initial power is accompanied by a long life, and only a small decrease of candle-power.

There are, then, three chief factors that enter into the real efficiency of the lamp, and these are: (1) life of the lamp; (2) mean candle-power; (3) mean power used. These factors depend on the way in which the lamps are manufactured, and when run at the pressure which the maker recommends, the resultant will give the true efficiency of the lamp, judged as a machine for converting electric energy into light.

Total energy consumed = mean watts × hours of life;

Total light given out = mean candle-power × hours of life;

and $\quad$ Efficiency $= \dfrac{\text{watt-hours}}{\text{candle-power hours}} \times$ hours of life.

If the lamp be run at other pressures than what is recommended by the manufacturer, whether slightly above or slightly below, different efficiencies will be obtained. The relations between working pressure and efficiency provide a subject for most interesting and elaborate calculations, and forms a neat example of the " maxima and minima " order of problems, because, as mentioned previously, running a lamp at a slightly higher pressure than what it is made for will give a large increase of light at the early part of its life, but the light will diminish rapidly, and the life of the lamp will be considerably shortened.

The following data give the results of experiments that have been made with the greatest care and every precaution by Messrs. Siemens and Halske, of Berlin, and they show in the most emphatic manner that those lamps that are run with a small initial power are not truly efficient, and that those which consume a smaller initial power are the most efficient in the long run.

TABULATION 30.

—	A.	B.	C.	D.	E.
Initial watts per candle-power	1·5	2·0	2·5	3·0	3·5
Final ,, ,,	3·65	5·24	5·47	5·27	4·32
Initial candle-power	16	16	16	16	16
Final ,, ...	5·6	5·2	6·6	8·4	12·5
Life in hours	45	200	450	1,000	1,000
Mean watts per candle-power...	2·85	3·69	4·21	4·04	3·85

Analysing this table, it may be remarked that lamps of type A take extremely small initial power—only 1·5 watts per candle-power, but their life is soon ended—only lasting 45 hours. Independent of the heavy cost of lamp renewals, the trouble and labour involved in the renewals would prohibit their use. Consider the work to be done in a building having, say, 1,000 lamps. A fresh lamp would have to be put in somewhere every three minutes. The mean watts per candle-power is certainly lower than what is required for a lamp of type E, and there is hence a saving of cost of power ; but an E lamp has 1,000 hours' life, and 22 lamps of the A type would have to be bought in order to get the same hours of burning as one lamp of the E type. Lamps used at this rate will soon bring up the expenses and nullify the saving that is effected by the small consumption of energy. Furthermore, type E after burning 1,000 hours

will still give a good light of 12·5 c.p., while absorbing the
same energy as at the start—namely, 56 watts ; but a lamp
of type A after burning 45 hours only gives 5·6 c.p.—a light
so poor that if the lamp were not at the end of its life it
would not be worth while using it.   An E lamp at the end
of its life of 1,000 hours consumes only 23 per cent. more
power per candle-power than at the commencement of its
life, whereas an A lamp at the end of its life of 45 hours
consumes 143 per cent. more power per candle-power than
at its commencement, the increase being more than six times
as much.   These figures show how extremely rapidly the
A lamps deteriorate, and that it is only in the early stage
of their life that they can claim to be efficient and economical in the true sense.

Having discussed the question of efficiency of lamps, so
far as the maker is concerned, we must now go further, and
find out the efficiency of the lamp from the consumer's point
of view.   The actual running cost of an incandescent lamp
may be composed of two factors, as follows : (1) the cost of
lamp renewals ; (2) the cost of energy consumed.

The first factor depends on the life of the lamp, and the
shorter this is the more will the cost be for lamp renewals.
The second factor depends on the maker's efficiency, or the
mean watts consumed per candle-power, and also on the
price of electricity.   The cost of lamps of, say, 16 c.p. may
be taken as fixed ; we may put down the cost of a lamp at·
2s., this price being the same for all lamps of 16 c.p., whatever their efficiency and whatever their life may be.

The price of electric energy will vary accordingly—some
places it is only 5d. per unit, at others as much as 1s. per
unit is charged.   When electricity is generated in private
establishments, such as in factories, the cost of production is
very low, and may be put down at, say, 2d. per unit, so the
running cost will be calculated over a range of from 2d. to

1s. per unit. From this it is easy to see that the price of electricity has a good deal to do in deciding how to run the lamp, because when the cost of supply is high, it may be best to slightly overrun the lamp. On the other hand, when the cost of supply is low, they can be run slightly under the normal pressure. This latter case particularly applies to private supply, such as a large factory, where the cost of generating electricity is extremely low.

Tabulation 31 shows the total running cost per candle-power hour of the five types of lamps given in Tabulation 30, calculated out for prices from 2d. to 1s. per unit, and based on Siemens and Halske's data in Tabulation 30.

In the case of type A, which takes an initial consumption of 1·5 watts per candle-power, the cost of lamp renewals is nearly twice as much as the cost of electric energy, even when at 1s. per unit, the exact figures being :

Lamp renewals = ·0626, energy = ·0342 ; total = ·0968.

TABULATION 31.

Price of supply unit in pence.	Total cost of 1 c.p. per hour in pence.				
	A.	B.	C.	D.	E.
2	·0683	·0207	·0138	·0100	·0093
3	·0711	·0244	·0180	·0141	·0132
4	·0740	·0281	·0222	·0181	·0170
5	·0768	·0318	·0264	·0222	·0209
6	·0797	·0355	·0307	·0262	·0247
7	·0825	·0392	·0352	·0302	·0286
8	·0854	·0429	·0391	·0342	·0324
9	·0882	·0466	·0433	·0383	·0363
10	·0911	·0503	·0475	·0424	·0401
11	·0939	·0539	·0517	·0464	·0440
12	·0968	·0576	·0559	·0504	·0478

With type B, taking 2·0 watts initially, the two components nearly balance when the price is 4d. ; then

Lamp renewals = ·0134, energy = ·0147 ; total = ·0281.

With type C, taking 2·5 watts initially, we have at the lowest price, 2d.,

Lamp renewals = ·00544, energy = ·00842 ; total = ·0138.

With type D, taking 3·0 watts initially, we have at same price,

Lamp renewals = ·00201, energy = ·00808 ; total = ·01.

With type E, taking 3·5 watts initially and at same price of 2d.,

Lamp renewals = ·00166, energy = ·0077 ; total = ·00936.

In all the above cases the lamp renewals remain constant throughout the range of prices per unit, and even when 1s. per unit is charged, type E costs less than one-half of type A.

When a lamp reaches the end of its life by its filament breaking, the light ceases, because the circuit is broken. Upon examining these spent lamps, it will be observed that the filament nearly always breaks from about ¼in. to ⅜in. below one of the platinum joints, occasionally they will break just before the bend ; the slightest fracture of the glass bulb of the lamp will lead to the destruction of the filament on account of air getting inside. One brake horse-power will run about 10 lamps of 16 c.p., and one kilowatt of electric energy will run 15 lamps at least, possibly 16.

Incandescent lamps are run in parallel at a constant pressure, the current increasing or decreasing according as the lamps in use increase or decrease. For example, if there be 50 lamps of 16 c.p. in use, worked at a pressure of 100 volts, they would probably take a current of about ·6 ampere each, or a total of 30 amperes, and this gives a consumption of 3,000 watts of electrical energy. Increase the number of lamps to 100 : the current taken will now be 60 amperes, and the electrical energy 6,000 watts, the lamps being in parallel. When the number is doubled there are double the number of paths for the current to traverse, consequently the total resistance of the circuit is halved, and since the pressure remains constant, therefore, by Ohm's law, the current becomes doubled. The greater the number of lamps the greater the current, and hence the greater the section of the mains.

The resistance of carbon decreases as the temperature rises, which is the exact opposite action of metals. This decrease is only very little for every degree C. increase of temperature, and the resistance of the filament of an incandescent lamp when alight is five-sixths the resistance when cold, so that in calculating the electrical energy consumed by a lamp, its resistance when hot must be taken. The resistance of a lamp of 16 c.p. when alight, and worked at a pressure of 100 volts, is about 160 ohms, and when cold, or not alight, it is about 190 ohms. The following data concerning the filaments of various lamps may prove interesting, all being worked at 105 volts pressure. The diameter is a little less than what it would be when the lamp was new, owing to its disintegration :

100 c.p.   ....   length, 28·6 cm., diameter, ·0635 cm.
 50 c.p.   ....   length, 26·0 cm., diameter, ·037 cm.
 16 c.p.   ....   length, 23·0 cm., diameter, ·013 cm.

When electric power is distributed over wide areas from one central source it is impossible to have the same potential difference everywhere, so that lamps that are situated near the source must inevitably be run at a greater pressure than those at the farther points. The Board of Trade allow a variation of 4 per cent. of the working pressure, so that there is a margin of 8 per cent. altogether, since the variation may be 4 per cent. below normal pressure or 4 per cent. above normal pressure. If the normal pressure is produced at a point or zone midway between the source of generation and the farthest point supplied, then a 16-c.p. lamp placed near the source will give about 20 c.p., and when placed at the farthest point, only about 12 c.p. ; this is when full advantage is taken of having the pressure at the source 4 per cent. too high, and at the farthest point 4 per cent. too low. Evidently the increase of candle-power is accompanied with a short life, and the decrease of candle-power with a long life. This maximum variation of pressure occurs when the maximum number of lamps are burning—that is, when the current in the distributing mains is at its greatest ; and as the current decreases and the number of lamps alight decreases, so the variation of pressure will decrease and approach nearer the normal pressure.

The following notes respecting the various appearance in colour which an incandescent lamp presents for various voltages may prove of interest.

The lamp experimented upon was made for 105 volts.

Pressure.	Appearance.
105 volts.........	full candle-power.
90 „	like an old lamp.
80 „	bright yellow, sharp filament.
70 „	dull „ „ „
60 „	much duller.
50 „	colour of sovereign, with reddish tinge.
40 „	„ red-hot iron.
30 „	very faint red glow.
20 „	faint red just perceptible.

The yellowish colour continues to about half the working voltage, then begins to turn reddish.

CHAPTER VI.

Low-Pressure System—Use of Three and Five Wires—
Loss in Mains—High-Pressure System—Alternate-Current
Working—Transformers—Insulation—Testing Circuits—
Traction Notes—Cost of Electric Energy.

The word " system " used in connection with the genera·
tion and distribution of electric power, is as much misused as
that other unfortunate word, " accumulator. A " secondary
battery ". no more accumulates electricity than a machine
gives out more energy than is put into it. The electric
energy put into it simply produces certain chemical reactions,
and so the electric energy is converted into chemical energy,
which latter can be reconverted back into electric energy
again when wanted by allowing the plates to regain their
original chemical condition. But there is no " accumulation "
going on ; in fact, it is rather the reverse, seeing that there
is a dead loss of above 25 per cent. of the electric energy put
into the cells. When the electric light was first exploited, in
the days of the mania of 1882, everything electrical was
somebody's "system," every arc lamp or dynamo put upon
the market was called some system or other of electric lighting ; and since the bulk of them were dismal failures, the
word got bad repute. No particular piece of apparatus constitutes a system, whether patented or not. Furthermore,
the dynamo was only invented once, crude, at first, certainly,
but gradually improved first in one way, then in another.
Then various well-defined types were evolved, as drum, disc,
and ring armatures ; horseshoe, double horseshoe, and multi-

polar field magnets; series, shunt, and compound winding.
By merely collecting the current we obtained alternating
currents of electricity; afterwards, inventive minds added
the commutator, and this gave us continuous currents of
electricity. Finally, coming close to the present time, two
or more pairs of separate collecting rings were keyed on the
shaft, each pair being connected to so many of the armature
bars, and thus we obtained two, three, or multiphase alternating currents. Again, the question of pressure was determined by the thickness and number of armature wires or
bars—armatures wound with a great number of thin wires
giving high-pressure currents, say at 1,000 volts, those wound
with a few thick turns giving low-pressure currents, say at
100 volts. Dynamos are not invented nowadays: like steam-
engines, they are designed, each dynamo-builder having his
own ideas of details of construction, the data for the same
having been collected by experience and actual experimental
tests, obtained at a great cost of time and money, in many
cases a score of differently-designed parts being tried and
condemned before the right one is found.

Broadly speaking, there are two chief and entirely distinct
methods of supplying and distributing electric power, and
which may with truth be called "Systems." These are
(1) the Low-Pressure Continuous-Current System, and (2) the
High-Pressure Alternating-Current System.

Low-Pressure System.

We will deal with this system first, as it is of less-
importance than the other. Low-pressure currents can
only be used to distribute electric power over very limited
areas, the reason being that when electricity is distributed
at a certain pressure from some central source there
always must be a gradual fall of pressure through the distributing mains, and this fall of pressure becomes greater

and greater as the distance from the source increases. A system of water service affords an exact analogy, as the pressure of the water becomes less the farther it travels from its supply source. From this it is evident that the fall of pressure will have a serious effect on the lamps. It was shown in Tabulation 28 that a slight fall of pressure of only 4 per cent. causes an enormous loss of light, as much as 25 per cent., so that while the lamps that are close by the source of generation burn brightly, and at their normal candle-power, the lamps will burn dimmer and dimmer as their position becomes farther away from this source. A fall of pressure being therefore inevitable, the best thing to do is to try to reduce this fall as much as possible. By Ohm's law we know that the fall of pressure along a wire is represented by the number of volts necessary to force a certain current through a certain resistance in ohms. Five hundred amperes flowing through a main cable 500 yards long and 1 in. diameter, and having a resistance of ·016 ohm, will signify a fall of pressure at the farther end of ·016 × 500 = 8 volts. The current density here being about 636 amperes per square inch, but this is only for one cable. The return cable must also be reckoned. This gives us a total drop of 16 volts, which is absorbed in sending the current through the cable—eight in the positive cable, and eight in the negative ; so that if the working pressure at the source be 100 volts, and a lamp be fixed 500 yards away, connected to the end of this cable, it will only receive 84 volts at its terminals, and naturally will be useless for illuminating purposes. This example will give an idea how great is the loss of pressure when heavy currents have to be carried any considerable distance.

There are several methods in use by which the low-pressure system is enabled to supply larger and more extended areas, such as by the adoption of what are called "feeders," the three-wire system, the five-wire system, etc., all these

being added to the simple parallel system. The simple
parallel system consists, as its name implies, of a pair of
mains, or a number of pairs, radiating out in various directions from the generating source, the sub-mains and branch
wires being tapped off along the route ; and since incandescent lamps only take a pressure of 110 volts at the most,
therefore the pressure at the generating station must be that,
or, say, two or three volts more, in order to allow for the
drop of pressure along the mains.   Suppose the station pressure is 112 volts when the full load is on, and that there is
a total fall of four volts between the station and the farthest
lamp burning ; now, when the load lightens and becomes,
say, one-half of the full load, this fall of pressure will be
only two volts, because, the current density in the distributing mains is only one-half of what it was before, and
hence the pressure required at the station will now be 110
volts.   This variation of pressure can easily be obtained by
a rheostat in the field of the dynamo, or else by varying the
speed of the engine.   A simple parallel system like this is
of very little use in distributing electrical power from central
stations, because the area it would supply would be too
limited in extent ; at the outside, it could only be used for
a radius of 200 yards.   This distance is no more than
would be suitable for a large manufactory or block of houses
where a private plant is put down.   The distance that a
certain current can be carried for a fixed fall of pressure
depends on the current density employed in the mains ;
when the density is high the distance is short, and when
low the distance is long.   The resistance of the mains is
constant in value whatever the current may be, so that as
the current increases in the main it requires more pressure
to force it through this resistance.   We arrive, then, at this
rule :

Distance is inversely $\propto$ current density.

This applies when the pressure at the source is constant in value, as it would be in any parallel system of distribution. The following distances in yards have been calculated out to show various falls of pressure in volts for various current densities per square inch sectional area of copper main.

TABULATION 32.

Current Density.	Fall of Pressure in Volts.							
	½	1	1½	2	2½	3	3½	4
	Distance in Yards.							
300	33	66	100	133	166	200	233	266
400	25	50	75	100	125	150	175	200
500	20	40	60	80	100	120	140	160
600	16	33	50	66	83	100	116	133
700	15	30	43	57	70	85	100	114
800	12	25	37	50	62	75	87	100·
900	11	22	33	44	55	66	78	89
1,000	10	20	30	40	50	60	70	80
1,100	9	18	27	36	45	54	63	72
·1,200	7	15	22	30	37	45	58	66

The range of fall of pressure is from one to four volts. The distance in yards signifies the actual distance of the lamp from the source of generation, and hence the fall of pressure is along a length of cable equal to twice this distance, because the length of both positive and negative cable must be taken into account. Taking the greatest distance shown—namely, that of 266 yards—there is here a fall of four volts, so that if 100 volts were given at the dynamo, only 96 would be received by the farthest lamp, fixed 266 yards away, and this is with an extremely low current density of 300 amperes per square inch. To employ such a low density would mean that an enormous weight of copper would be sunk in mains and branch mains, and the expense of this mass of copper would be extremely·

great. To attempt to carry these mains any farther away
would make the fall of pressure so much that the lighting
would be perfectly worthless, because if the fall of pressure was doubled, and made eight volts, it would only
give a working radius of 532 yards, and so it is clear
that it is practically impossible to distribute power over
any wide areas, like a town of five or six square miles,
by using a simple parallel low-pressure current of only
100 or so volts.

Network Feeders.—In order to equalise the pressure over a
network of distributing mains as much as possible, and to
enable a farther distance to be supplied without any increase
of the fall of pressure in these mains, the method of trunk
mains was adopted, these trunk mains being used solely
and simply for conveying the current to several determined
centres in the network. No branch wires are tapped off
these trunk mains, and no lamps are run off them, so that
they serve as "feeders," and are known by that name. It
is evident that the more "feeders" there are, the more
will the pressure be equalised over the system, because
there will be more central points, at all of which the
pressure will be maintained constant and the same. At
the end of each pair of "feeders"—that is, at each "feeding
centre"—a thin pair of "pilot" wires are taken back to the
dynamos, so as to indicate what the pressure is at the centre.
As a practical illustration, we may take it that 112 volts is
the pressure that must be maintained constant at the feeding
centres, and at full load; this may mean, say, 118 volts at
the terminals of the dynamos : for the nearer centres a little
less, for the farther centres a little more, when all the centres
have different distances. As the current flowing in each
feeder increases or decreases, so a greater or less pressure
will be required at the dynamo terminals in order to always
maintain the 112 volts pressure at the feeding centre.

Taking as six volts the total fall of pressure along the feeder at full load, there will only be required 115 volts at the dynamo when the load in that feeder is reduced to one-half. By adding a number of " feeders " to the simple parallel system, the radius distance of the supply mains can be increased to double, treble, or even more. The greater the length of feeders supplied, and the more feeding centres

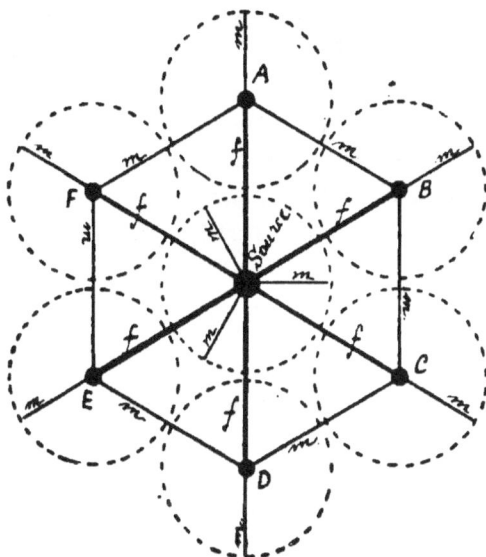

FIG. 31.

there are in the network of mains, the more equalised will be the pressure.

No low-pressure systems of distribution are carried out without employing feeders in a judicious manner, and as they are solely for the purpose of giving better regulation to the supply mains, the money sunk in them becomes an additional expense. It is found that in a well-designed distribution system the money spent on feeders should be

U

between 30 per cent. and 50 per cent. of the money spent on
the distributing mains, according to the nature of the area
to be supplied.

Fig. 31 gives an idea of the way feeders are used. Where
A, B, C, D, E, F are the six feeding centres, and f the
"feeders," the distributing mains denoted by m, radiate out
from each centre, and those nearest each other are joined
together, and so make a circuit round the source ; only the
positive wires are shown, to make the diagram clearer. Of
course the position of the centres and their number and
proximity to each other are determined by the nature and
extent of the district, and the diagram given is merely to
show how they are employed, it being almost impossible to
get a district where the distributing mains and feeding
centres can be arranged perfectly symmetrically, as in the
case of Fig. 31.

Use of Three and Five Wires.

If the current that is flowing is reduced to one-
half of its value, whilst at the same time the pressure
is doubled, we still have the same electrical power ; and
generally the current can be reduced by as many times as
the pressure is increased without altering the value of the
product which represents the power generated. We showed
that a main cable of 1 in. diameter and 500 yards long had
a resistance of about ·016 of an ohm, and that with a current
of 500 amperes flowing through, the fall of pressure throughout its length was eight volts, so that if the pressure at the
generating end of the cable be 100 volts the fall is then
8 per cent. But suppose we work our lamps with a current
of only half that value, 250 amperes, and at a pressure of
200 volts, the power is the same ; but instead of our lamps
being fixed in simple parallel they must now be arranged in
parallel rows of two in series—that is, two lamps in series

must now be placed across the positive and negative mains so as to absorb the 200 volts—because the lamps are only made for 100 volts pressure. The same current passes through the two lamps that are in series, so that the total current used is obtained by multiplying this current by one-half of the number of lamps in use. In the simple parallel arrangement, the total current is obtained by multiplying the current through the one lamp by the number of lamps there are. From this, we see that for the same number of lamps we only require one-half the current when the pressure is doubled.

If our current is now 250 amperes, the sectional area of the cable can be reduced correspondingly, and this will give us the same current density, the resistance is doubled, and so we have the same fall of pressure—namely, eight volts ; but the working pressure is now 200 volts, and so we obtain a fall of only 4 per cent. instead of 8 per cent., as with 100 volts.

Reckoning in the negative cable as well, there is a total fall of pressure of 16 volts in both cases, but on the 100-volt circuit each lamp received only $100 - 16 = 84$ volts, whereas on the 200-volt circuit each lamp receives $(200 - 16) \div 2 = 92$ volts. Therefore, keeping the percentage of loss constant—that is, the loss for each 100-volt lamp—doubling the pressure will allow the current to be carried double the distance, or

Distance is proportional to Pressure.

The above fact furnishes us at once with a means for extending the distributing area of a system. The three-wire system of distribution enables us to use 200 volts, the third wire being called the " neutral wire," because when there is a balance of load of each side of it—that is, when there are as many lamps burning between the positive and

neutral wire as between the negative and neutral wire—
there is no current flowing in it.   Fig. 32 shows the connections of the three wires.

The dynamo is wound for a constant pressure of 200 volts,
and the pressure can be regulated by a rheostat in the shunt
coils, for compensation for fall in the circuit.   The third wire,

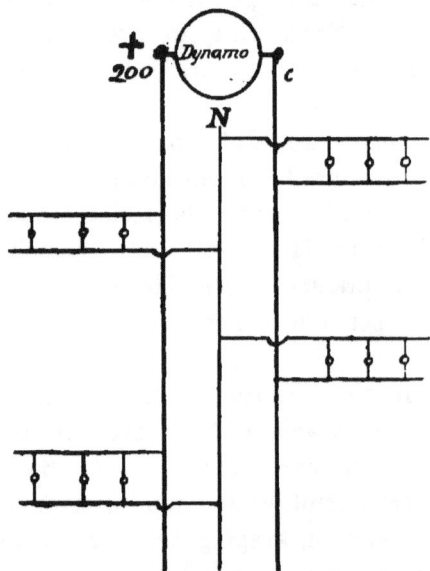

FIG. 32.

signified by N, is placed between the positive and negative
wires, but does not return to the dynamo.   Service mains for
100 volts can be taken off either + and N or N and −,
because the + is at 200 volts pressure and − at 0.   Of
course the circuit can only be completed when there are
lamps on *both* pairs of mains, as the lamps of one pair are in
series with the lamps of the other pair.   Fig. 32 shows four
tappings of service mains, two being taken from + and N,

and two from N and −. By having this neutral wire, N should one lamp go out the others will not be thrown out of circuit, whereas if the lamps were fixed two in series straight across + and −, then one of the pair going out would make its fellow go out.

Since there is a current flowing through the neutral wire only when there is an *unequality* of load, it is usual in practice to make its sectional area not much more than one-half what either of the other two may be; in some cases it is made only one-third. There is, therefore, a great saving of copper in the three-wire system as compared with the two-wire system. Allowing a two-wire system to have cables of 1in. diameter, the total sectional area of copper employed is 1·5708 square inches.

With the three wire system, we have first the sectional area of the two mains, + and −, reduced to one-half, because one-half of the current is used; but at 200 volts, the percentage of fall of pressure for equal distances is only one-half, so that to bring it up to the same as the two-wire system, we can double the current density, assuming that this will not overheat the wire, and this can be done by reducing the sectional area again by one-half, so that the two mains have now a total area of one-quarter of those on the 100-volt circuit, or ·3927 square inch, each main being ·5in. diameter. The third, or neutral, wire having one-half the area of either main, will have ·098 square inch; so the three wires will have a total of ·3927 + ·098 = ·4907 square inch, or very nearly half a square inch. Hence we have

Two-wire system = 1·5 square inch ⎫
Three-wire system = ·5 square inch ⎭ approximately,

giving the latter a saving of 66 per cent. in weight.

It must be remembered that this is under the conditions of same distance and same percentage of fall of pressure.

By allowing the three-wire system to go to twice the radius distance, keeping the percentage of fall constant, there would then be a saving of 33 per cent. in the weight of copper used.

Fig. 33 shows how "feeders" are supplied to a three-wire system, there being three feeding centres, A, B, and C. The

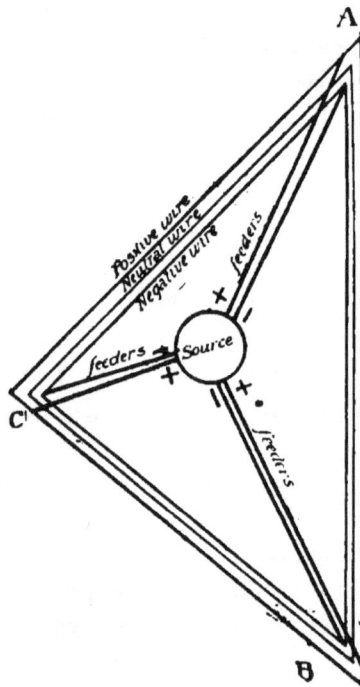

FIG. 33.

outer cable is positive, the inner one negative, whilst the third, or neutral, wire is in the middle. Both positive and negative feeders are shown, the positive feeders being connected to the outer wire and the negative feeders to the inner wire.

Fig. 34 shows a number of dynamos working into a three-

wire system, and also how secondary or storage cells are used
in conjunction with the dynamos. The method of connecting
the dynamos and cells is that usually adopted in central-

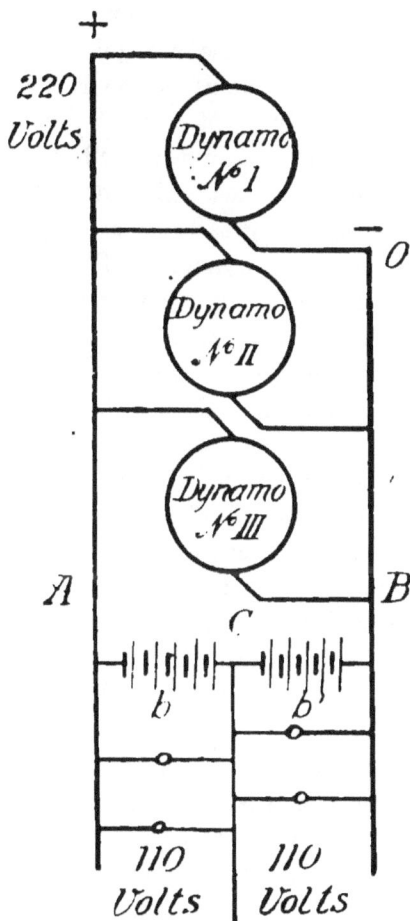

FIG. 34.

station practice for three-wire systems. As the load increases,
another dynamo is switched into parallel, as on the two-wire
system. There are three machines shown in parallel in

Fig. 34, all working at a pressure of 220 volts. The cells serve two purposes—first, they steady the supply, discharging a current when the load makes heavy demands, thus easing the sudden strain on the dynamos ; and absorbing current from the dynamos when the load suddenly falls, thus replenishing the electric energy in the cells, so they may be looked upon somewhat as a " reservoir." Two batteries of cells are used, b and b'. These are joined in series across the positive and negative mains, A and B, the third wire, C, being run from the junction of b and b'; so when the load on the + side is greater than the load on the − side, the battery b assists the dynamos. On the other hand, when the load on the − side is greater than that on the + side, then battery b' assists the dynamos. So far, the third wire is not connected up to the dynamos in any way, and in the case where feeders are used, as in Fig. 33, the third wire is laid down from the feeding centres, thus saving a considerable weight of copper by not taking it back to the generating source.

The original use of the third wire, as invented by Dr. John Hopkinson in 1882, wàs obtained by coupling together two similar dynamos in series, the + and − cables being connected to the outer + and − terminals of the set, while the third or middle wire was connected to the junction of the inner + and − terminals. The advantage of thus taking the third wire back to the dynamos is that it enables the surplus current to return to the dynamos when the load is greater on one side of the system than on the other side. It need hardly be said that in connecting up làmps on a three-wire system it is necessary, as far as actual practice will permit, to arrange the lamp circuits so that the whole of the supply or load is às nearly equally divided as possible— that is, to have as much current taken off the + and third wire as off the − and third wire. When the load on the + side is greater, say, than that on the − side, the difference between the two currents is called the " differential current," and this current will flow along the third wire back to the dynamo on the + side, because that dynamo is working at a greater load than the other. If the − load were greater than the + load the differential current would flow back along the third wire to the dynamo on the − side, and when the loads on both sides are exactly equal then there is a balance and no current flows along the third wire, because the + and − sides are then just as if there were no third wire and they were in simple series. From this the third wire is often called the " balancing wire."

The next development of the parallel systems is found in the five-wire system, and as the three-wire permitted double the pressure that was used in the simple parallel, and hence an extension of area, so the five-wire system permits double the pressure that is used in the three-wire system, and hence a still further extension of area. In this system the dynamos work at, say, 400 volts, three wires being placed between the two outer wires, thus making five wires altogether. Lamp circuits requiring 100 volts or so can thus be taken off any two neighbouring wires out of the five ; if 200 volts is required, the circuit is taken off any two alternate neighbouring wires ; if 300 volts be required, it is taken off, say, the outer + main and the fourth wire and so on. There are thus four circuits of 100 volts in the five wires taken respectively off the + main and the second, the second and the third, the third and the fourth, and the fourth and the − main. Every circuit must have some lamps connected on, and these should as far as possible be distributed equally between the four circuits so as to obtain an equalised load, so in reality these four circuits are all in series, and so make up the 400 odd volts that exist between the two outer cables.

Parallel systems are mostly designed for 110 volts pressure at the distributing centres, because 105-volt lamps are mostly used, and the other five volts is a margin for fall of pressure. This explains why in Fig. 34 the dynamos are of 220 volts, and the two circuits of 110 volts each ; hence in the five-wire system the dynamos are worked usually at 440 volts, giving four circuits of 110 volts each. In Fig. 35 the five wires are not all kept together throughout the supply district, but the diagram is drawn to represent the method of supplying two districts that are close together, and both a little distance away from the central station. Two dynamos of 440 volts pressure are shown in parallel—A is the positive cable, and B is the negative ; and on account of the pressure being much higher than a simple parallel system, the current flowing through them is much less, and consequently the size and weight of the cables are considerably less, which is a matter of paramount importance when distance has to be considered. When they reach the two districts to be supplied, a third wire, C, is introduced, the + cable and the third wire, C, supplying one district, whilst the negative cable and the third wire, C, supply the other district. We have now a pressure of 220 volts in each district, and again a third wire is added, one to each district, thus making a total of five wires in all, each being at different pressure. In the left-hand district two lamp circuits can now be taken, each of 110 volts. Circuit b being taken from the middle wire and A, A is at a pressure of 440 volts, the middle wire will have a pressure of 330 volts, and the difference is 110 volts. Similarly, circuit a takes the difference between the middle wire and C, then $330 - 220 = 110$ volts as before.

Passing along to the right-hand district, we can in a similar way take off two more circuits, c and d, each having 110 volts, the last circuit, d, taking the difference between 110 and

0 = 110, so that the circuits and their lamps are all in series between A and B, and thus have a working pressure of

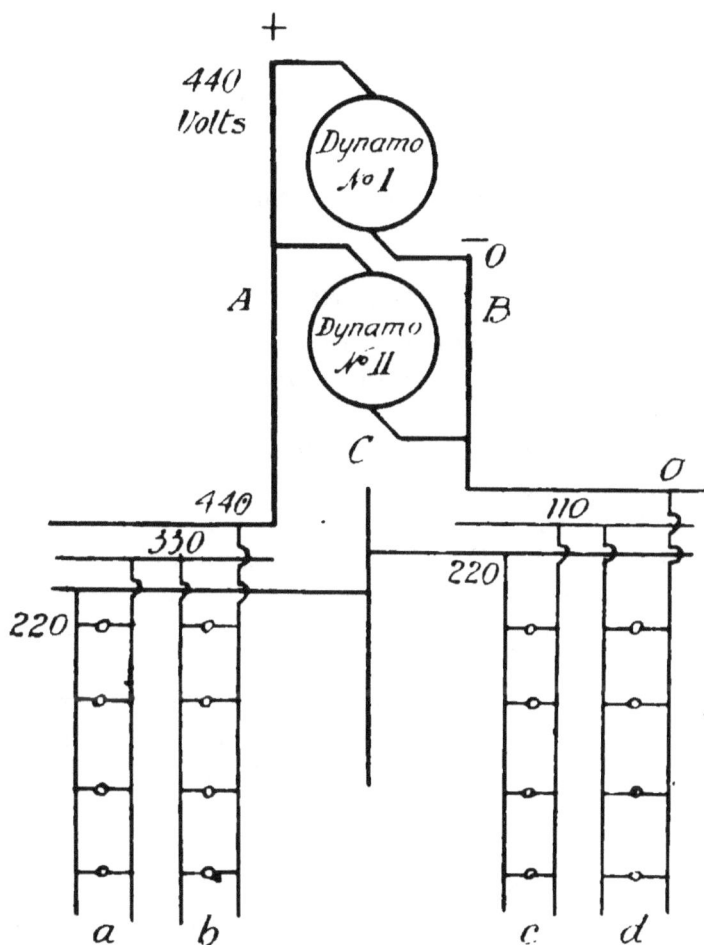

FIG. 35.

440 volts. Great flexibility of distribution can thus be obtained from the use of a five-wire system : it is certainly complicated, particularly when the five wires are carried up

to the lamp circuits, and on that account, together with the
extra expense in dealing with such complications, is at a disadvantage as compared with simple parallel wires of a high-
pressure alternating current.

When feeders are used in the five-wire system they are
connected to the two outer wires, there being three intermediate wires, so that in Fig. 33, if the middle wire were
replaced by three wires, and the pressure of the machines
changed from 200 to 400 volts, that would give an illustration
of this method of supply.

Loss in Mains.

The loss or waste of electric energy along a pair of mains is
a very different thing from the mere loss or fall of pressure,
and careful distinction must be made between the two,
because the latter depends simply on the current and the
resistance of the mains, denoted by CR, while the former
depends on the square of the current and the resistance,
denoted by $C^2 R$. We will now investigate what results are
obtained when a pair of mains, or feeders, undergo various
conditions of working whilst the electric power received by
them remains the same, it being assumed that none of the
electric power is used along the mains, but the whole of what
is transmitted is utilised at the far end of the mains, so that
the cables under consideration may be looked upon really as
" feeders."

The following will be assumed as the initial or normal
condition of things : The feeders have a sectional area of one
square inch each, a distance of 200 yards (200 yards positive
and 200 yards negative), and a current of 400 amperes flows
through at an initial pressure of 100 volts; hence the current
density is 400 amperes per square inch, and the electrical
power delivered by the dynamo into its end of the feeder is
$100 \times 400 = 40,000$ watts, or 40 kilowatts.   The resistance

of such a pair for the above distance of 200 yards may be put down at ·01 of an ohm—that is, one-hundredth of an ohm, being at the rate of ·044 ohm per statute mile when taken at the temperature of 60deg. F., or 15·5deg. C. The increase of resistance due to the current may be neglected, because it is only very small with a low current density of 400. The total weight of copper in both feeders for this distance is 2·08 tons, or 4,660lb. To force 400 amperes through ·01 of an ohm resistance requires ·01 × 400 = 4 volts, so that the pressure at the farther end of the feeders will be 96 volts, two volts being lost in the positive feeder and two in the negative, hence the total fall of pressure is 4 per cent. We can now calculate the loss of energy in the feeders by multiplying the fall of pressure by the current. This is the same thing as multiplying the line resistance by the square of the current, because $C^2 r = C e$, where $e = C r =$ fall of volts. Applying this to the above case, we have $4 × 400$, or $(400)^2 × ·01 = 1,600$ watts. The total energy delivered to the feeders is 40,000 watts, the quantity received at the end of the line is 38,400 watts, so that the loss of energy = 4 per cent. Doubling the pressure and using the same current density, the same power can be transmitted double the distance for the same percentage of fall of pressure or loss of energy; so with 200 volts pressure, the feeders need only have a sectional area of ·5 square inch each, and the current of 200 amperes can thus be driven 400 yards.

It is evidently very clear that when the percentage of fall of pressure or loss of energy must be confined to certain low limits, it is absolutely necessary to employ small currents of high pressure in order to transmit or distribute electric power over long distances or scattered districts.

To show at a glance how the loss varies under different conditions, Tabulation 33 is prepared, the electric power delivered in every case remaining constant—viz., 40 kilo-watts. The three most important variables being (1) pres-sures (2) distance, (3) current density, when these are known the rest can be found.

TABULATION 33.

Pressure in volts.	Distance in yards.	Current density per square inch.	Area of cable in square inch.	Weight of copper in tons.	Total pressure lost.	Percentage of loss.	
						Pressure.	Energy.
100	200	400	1·00	2·08	4	4	4
200	200	200	1·00	2 08	2	1	1
200	400	200	1·00	4·16	4	2	2
200	200	400	·50	1·04	4	2	2
200	400	400	·50	2·08	8	4	4
400	200	400	·25	0·52	4	1	1
400	400	400	·25	1·04	8	2	2
400	800	400	·25	2·08	16	4	4
800	800	400	·125	1·04	16	2	2
800	1,600	400	·125	2·08	32	4	4
1,600	1,600	400	·0625	1·04	32	2	2
1,600	3,200	400	·0625	2·08	64	4	4

Comparing together the first and the last, it is seen how enormously the distance can be increased by using high-pressure currents, the loss being the same, the weight of copper being the same.

From these figures several useful rules can be obtained :

(1) With current density and distance constant, percentage of loss is ∝ inversely to pressure.

(2) With current density and percentage of loss constant, distance is ∝ to pressure.

(3) With current density and pressure constant, percentage of loss ∝ distance.

(4) With distance and percentage of loss constant, current density ∝ pressure.

(5) With distance and pressure constant, percentage of loss ∝ current density.

(6) With pressure and percentage of loss constant, distance is ∝ inversely to current density.

TABULATION 34.

Conditions of Supply.	Two-wire.		Three-wire.		Five-wire.
Pressure in volts	100	...	200	...	400
Current in amperes	400	...	200	...	100
Total section of copper in sq. in. ...	2·0	...	1·25	...	·75
Total weight of copper in tons	2·08	...	1·30	...	·78
Percentage of loss of pressure	4·0	...	2·0	...	1·0
Percentage of loss of energy	4·0	...	2·0	...	1·0
Percentage of copper used............	100·0	...	62·5	..	37·5
Percentage of copper used .... with 4% loss constant	100·0	...	31·25	...	18·75
Supply distance in yards	200	...	400	...	800

A summary of the three methods of low-pressure distribution is given above in Tabulation 34, in each case the power received from the generating source being 40,000 watts, none being tapped off along the line, so that the whole of what is delivered is utilised at the far end, the distance served being 200 yards from the generating source. The last row of figures in the tabulation gives the increased distance the power can be sent, when the percentage of fall of pressure or loss of energy is kept constant.

High-Pressure System.

We may now pass on to consider the second general system of distributing electric power—namely, by means of high-pressure alternating currents. From the preceding figures it is seen that, generally, the higher the pressure employed the greater in proportion is the distance a given power can be transmitted and distributed, keeping the loss of energy constant; also, that for a given distance and loss of energy the weight of copper required will vary inversely as the square of the pressure employed.

This last law is, however, only theoretical, on account of making the loss constant for each case. The usual pressure used on high-pressure circuits is about 2,000 volts. Suppose we use 20 tons weight of copper to transmit 40 kilowatts a distance of 2,000 yards, with a loss of 40 per cent. or 16,000 watts, when a pressure of 100 volts is used. Then, if we used a high pressure of 2,000 volts, we should only require a current one-twentieth the amount; hence the sectional area, and consequently the weight of the new conductors, need only be one-twentieth also—that is, one ton. But the loss of energy on the line is obtained by the expression $C^2 R$, so that although the resistance of the conductors increases while the current decreases, it must be remembered that the loss depends on the square of the current. Comparing the two losses, we have

$$\text{with} \quad 100 \text{ volts, } C^2 R = 400^2 \times \cdot 1 = 16,000 \text{ watts ;}$$
$$\text{,, } \quad 2,000 \quad \text{,, } \quad C^2 R = \quad 20^2 \times 2 = \quad 800 \quad \text{,,}$$

so that the loss in the second case is only one-twentieth of the loss in the former. But the law states that the loss must be constant, so that the resistance of the 2,000-volt circuit must be either increased 20 times, or the resistance of the 100-volt circuit must be decreased 20 times ; and this is where the result is absurd in a practical sense, because it would signify in the former a current density of 8,000 amperes per square inch, and in the latter case a density of only 20 amperes : the first would burn up the wires, while the second would be like burying money in the ground. The result, however, would be that the weight of the copper in the 2,000-volt circuit would be still further reduced to one-twentieth again, and as it has already been reduced to one-twentieth before on account of the diminished current, this makes, finally, one-twentieth of one-twentieth, or one four-hundredth, so that using 20 tons with 100 volts, we should only require $20 \div 400 = \frac{1}{20}$ of a ton with 2,000 volts, thus exemplifying the theoretical law given above.

Neglecting to keep the loss of energy constant, and, instead, keeping the current density constant, we can then reach a practical solution, and this is, that the weight of copper used is inversely proportional to the pressure used, while at the same time the loss of energy becomes decreased in proportion as the pressure used rises. There is thus a great saving to be effected in the use of high-pressure currents, for not only is the weight of copper decreased, but the loss also. Applying this practical law, we find that the cost with a 100-volt circuit would be, say, $20 \times 40 = £800$ for copper, whilst with 2,000-volt circuit the cost would be only $1 \times 40 = £40$, and the loss of energy only one-twentieth of the loss in the former instance.

Although these advantages are very great and of a substantial nature, so far as concerns financial matters, it must not be thought that there are no drawbacks to the use of the high-pressure system. The greatest of these, and the one most troublesome to deal with, is want of reliable insulation. The greater the pressure at which an electric current is driven along a conductor, the more difficult is it to confine it to the conductor and so prevent leakage. It is comparatively easy to well insulate a cable that is worked at 100 or 200 volts pressure, and the indiarubber covering need only be fairly thin, and hence inexpensive ; but when dealing with 2,000 volts, heavy and very expensive covering is absolutely necessary, so that, roughly speaking, a 100-volt main is composed of a thick copper conductor with a thin covering, whilst the 2,000-volt main is a thin wire with a covering composed almost entirely of guttapercha and indiarubber, and having a depth perhaps greater than the diameter of the conductor. The cost of a cable is composed of two factors : (1) the cost of the mass of copper therein, (2) the cost of the insulating covering on the copper. In the case of low-pressure distributing cables, it is evident that the cost of copper as compared with the cost of insulation is high, while with high-pressure cables the cost of copper, as compared with that of insulation, is low, so that the ratio between the two varies greatly and inversely. For example, if we transmit 40 kilowatts of electric power at 100 volts one mile, we might use a cable costing £1,800 per mile, while if we transmit the same power at 2,000 volts pressure, we should use a cable which would only cost about £300 per mile. In the latter case the weight of copper is only about one-twentieth the weight used in the former case, and yet the total cost of the latter cable is one-sixth of the former.

Fig. 36 shows how connections are made on a high-pressure alternating-current system. On an alternator, as already explained, there is no commutator, and the two ends of the armature wires are led to two rings, or " collectors," fixed in the armature shaft, each well insulated from the shaft and from each other. A couple of brushes are fixed on each collector, and the two circuit wires are led from the two collectors, as shown in the diagram. One is marked + and the other − in order to distinguish the two, although the polarity of the machine is constantly changing from + to − and back again. The two line wires are of small diameter, but thickly insulated, as only small currents are used. This thin wire and any tappings that may be taken off in parallel is named the " primary " circuit, and the two terminals of the fine wire or primary coil of a transformer are joined up to the primary circuit. This coil is indicated by the thin zigzag lines ; the thick zigzag lines drawn opposite it denote the thick or secondary coil of the transformer, and its terminals are therefore connected on to the thick wires of the

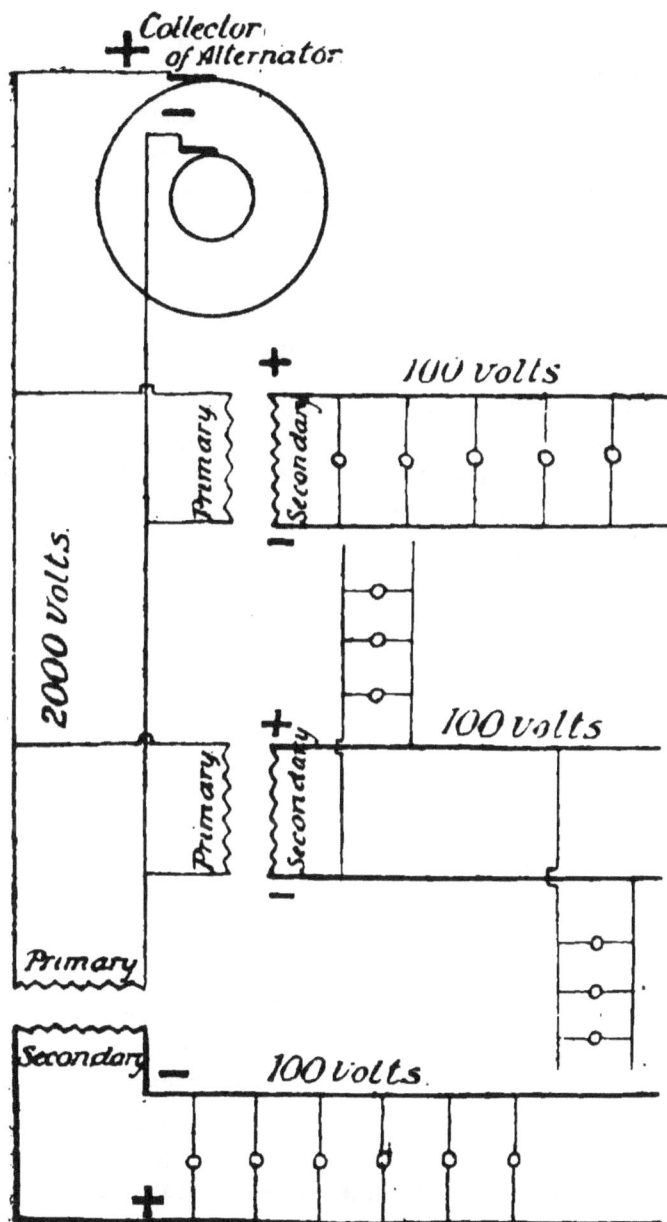

Collector
of Alternator

100 volts

100 volts

100 volts

2000 volts.

Primary

Secondary

Primary

Secondary

Primary

Secondary

FIG. 36.

secondary circuit, or lamp circuit, from which the lamps are run in the usual parallel way.

The pressure in the primary coil of the transformer is that of the primary circuit—namely, 2,000 volts ; while the pressure in the secondary coil and secondary circuit is only 100 volts, so that the function of the transformer is to receive a small current of high pressure and to send out a large current of low pressure.

Alternate-Current Working.

In entering upon this subject the reader must be prepared to tolerate a little elementary mathematics, since the theory of alternating currents cannot be understood without. The whole subject, indeed, is one that calls for a fair mathematical knowledge, and every day the necessity of the study and examination of this branch of electricity becomes more and more compulsory to those who wish to keep pace with the development of electrical science. In direct-current work everything is so simple that those who have only had experience in that branch find themselves quite at sea when called upon to operate alternate-current plant, and this is not to be wondered at. A direct-current engineer has only to multiply his ammeter reading by the voltage of the circuit to obtain the electrical power that is going out ; the alternate-current engineer cannot do this. If an alternator is working at 2,000 volts pressure, and the current in the primary circuit is 40 amperes, the power given out is not $(2,000 \times 40)$ watts. Similarly, if the alternating volts be divided by the resistance in ohms of the primary circuit, the quotient will not give the primary current in amperes. It is on this account that it is said that Ohm's law is not true for alternating currents, but it is perfectly true if applied in the right way, and all the other causes and effects are taken into account. There are more E.M.F.'s than one, and there are other obstructions to·

the current besides the mere metallic resistance of the con-
·ductors ; they ignore the effects of self-induction and the
lagging of the current behind the impressed E.M.F. Only
a few of the rudimentary equations can be given here, and
those who wish to enter into the subject should consult
Dr. Fleming's work, entitled " The Alternate-Current Transformer in Theory and Practice"—the theoretical part is
in Vol. I.

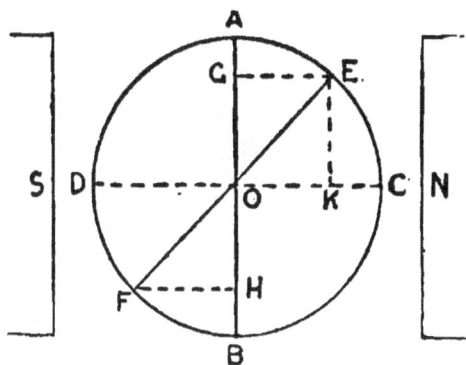

FIG. 37.

The nature of an alternating current has already been explained, and on referring back to page 170, Fig. 20 shows the
two curves—the E.M.F., and the current that lags behind. It
is necessary in dealing with this kind of work to know the
value of both the E.M.F. and current at any instant—that is,
at any stage of the periodic rise and fall. Suppose we have
a coil of wire, as in Fig. 37, capable of rotation, and placed
in a magnetic field whose strength is H ; let the area of the
coil be A, and let it make one complete revolution in the
time T, then at any instant t the number of lines of force
embraced by the coil are

$$N = H A \cos \frac{2 \pi t}{T},$$

where the fraction $\frac{t}{T}$ determines the angle in degrees through which the coil has turned from the vertical.

The slope of the curve at any point is obtained by finding the tangent of the angle which the curve at that point makes with the horizontal; evidently the angle cannot be greater than 90deg., because then the slope of the curve is at right angles to the horizontal—that is, it occupies a perpendicular position. The maximum value of the tangent is $N \frac{2\pi}{T}$, and when the cosine is diminishing, then the sine curve has a positive value, and *vice versâ*. If we suppose the number of lines embraced to be decreasing, we shall find that the time rate of change, or the E.M.F., E, produced, to be

$$E = \frac{2\pi}{T} A H \sin \frac{2\pi t}{T};$$

the expression $\frac{2\pi}{T} A \underset{\sim}{H}$ is, therefore, the maximum value of the time rate of change of N, and it is also interesting to note that if we have any quantity, N, undergoing variation according to a sine law, then the maximum value of rate of change will be

$$\frac{2\pi}{T} N.$$

The pulsating waves of an alternating current so closely resemble a curve of sines that the non-mathematical reader should establish for himself the following most important facts :

(1) The mean ordinate of a sine curve between values 0 and π is $\frac{2}{\pi}$ or ·637 M, where M represents the maximum ordinate. In Fig. 38 the area enclosed by the sine curve, O M T, is obtained by multiplying the maximum height, M,

by the base, O T, and then by ·637. The area of the curve between the limits O and T is thus equal to the rectangular area, O A B T, since the ordinate O A is ·637 of M. Since the number ·637 is not far from ·666, hence we may say that the time average of the fluctuating current is, roughly, about two-thirds of the maximum value of the current.

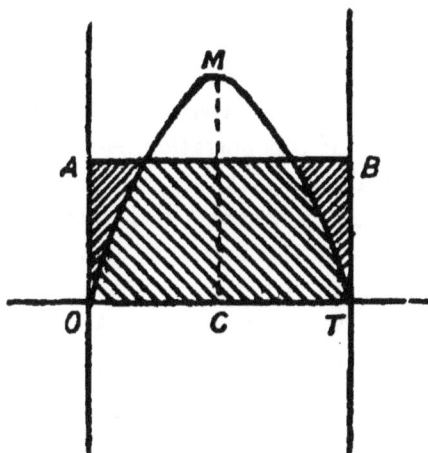

FIG. 38.

(2) In measuring an alternating current it is the square of the current that must be considered; hence the average given in (1) is not the true measure of the mean current, but it will be the square root of the mean square, expressed by $\sqrt{\dfrac{M^2}{2}}$ or $\dfrac{M}{\sqrt{2}}$, because the square of the maximum value is M^2, and the mean square is $\dfrac{M^2}{2}$; so that we have two mean values, the first being ·637 of the maximum, or the simple arithmetical mean, and the second being $\dfrac{1}{\sqrt{2}}$, or ·707 of the maximum, or the square root of the mean square. This last expression is now written thus, $\sqrt{\text{mean sq.}}$ for terseness.

The difference between ·637 and ·707 is ·07, which thus shows that the simple mean is 10 per cent. below $\sqrt{\text{mean sq.}}$ Draw two sine curves, one of which is retarded by a small angle, ϕ, behind the other ; multiply their ordinates together at every point, and take the mean value ; increase the retardation, ϕ, and repeat the operation, and we shall find that we get a smaller result. Make $\phi = 90$deg., so that the second curve commences at a quarter of a period behind the first curve, and we shall find that our result is now zero, and no work is done because the positive products exactly equal the negative products. If this operation be repeated for several values of ϕ, we shall find that the average value for a given value of ϕ is A B cos ϕ, where A and B are respectively the maximum ordinates of the two sine curves under examination. In an alternate-current circuit the current would be equally in phase with the impressed E.M.F., if it were not for the disturbing effect of self-induction, which causes the current to be retarded, or to lag behind the impressed E.M.F. The following experiment illustrates how self-induction produces erroneous readings when ordinary measuring instruments which are made for direct-current work are used for alternate-current work.

A horizontal coil, having a resistance of 1,158 ohms and consisting of 1,150 convolutions, has suspended at its centre a smaller coil, having a resistance of 65 ohms and 2,150 convolutions, the axis of the smaller coil being at right angles to that of the larger. The ends of the two coils are connected in series, and are then first joined up to a source of . steady E.M.F., and the deflection of the small coil noted ; the circuit is then broken, and the ends of the coils are joined up to a source of alternating E.M.F. of equal amount, when it is found that the deflection is much less ; in fact, in one case, when the " frequency," or " periodicity," of the current was 260 per second, the deflection was 15 per cent. lower than the deflection obtained from a direct E.M.F. of equal amount. The accompanying curve shown in Fig. 39 shows the percentage of error observed when a voltmeter containing iron, and constructed for direct currents and steady E.M.F.'s, was used with alternating E.M.F.'s of various periodicity. With a periodicity of 133 per second, the reading on the voltmeter was no less than 26 per cent. lower than what it would be if the E.M.F. were steady, instead of being of an alternating character. The above experimental fact teaches us one very important lesson, and this is : that wherever there is any iron present in an alternating-current circuit, it will create self-inductive disturbances. Therefore, most instruments for measuring alternating currents, whether ammeters, voltmeters, or wattmeters, must be constructed free from iron.

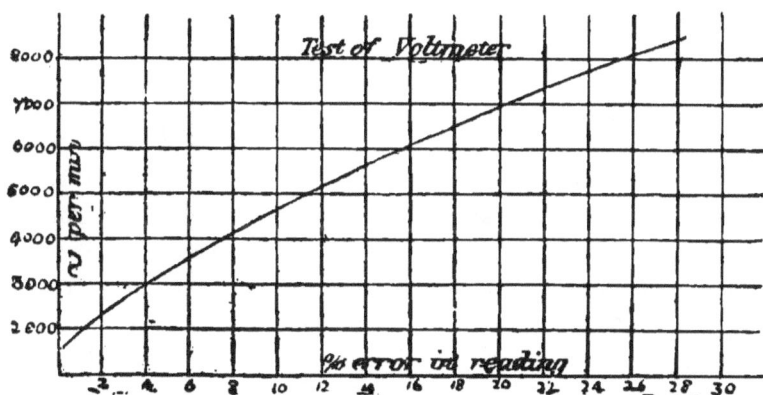

FIG. 39.

An alternating-current circuit may be one of two kinds— inductive or non-inductive. We will consider the latter kind first, as it is the simpler : a non-inductive circuit signifies one that has no self-induction, consequently the only obstruction to the current is the actual ohmic resistance of the conductors.

To measure the alternating current, we should take an electro-dynamometer which will measure the $\sqrt{\text{mean square}}$ or $\dfrac{I}{\sqrt{2}}$, where I signifies the maximum value of the current ; similarly, to measure the alternating volts, we should use a Cardew voltmeter, which will measure $\sqrt{\text{mean square}}$ of the impressed E.M.F., or $\dfrac{E}{\sqrt{2}}$, where E signifies the maximum value of the impressed E.M.F.   Having now obtained both the current and the volts, the product of the two readings will give us $\dfrac{E\,I}{\sqrt{2}\,\sqrt{2}} = \dfrac{E\,I}{2}$, and this represents the mean power consumed.

The above is true when both the E.M.F. and current are in step or in phase, but suppose that the current curve is retarded, and lags 60deg. behind the E.M.F. curve, then we shall have

$$\text{Mean power} = \frac{E\,I}{2} \times \cos 60\text{deg.} = \frac{E\,I}{4}.$$

Suppose we have a number of forces acting on a point, it can easily be shown that if lines are drawn parallel to these forces, and proportional in length to the forces represented, and to some fixed scale, so that in direction and magnitude they may symbolise the several forces ; then, if all the forces are in equilibrium, the figure obtained will be a closed polygon, and if they are not in equilibrium, the figure will be open. For example, suppose in the graphical view given in Fig. 40 A, B, C, D, E, F, G, H represent such a polygon, obtained from a number of forces, and open at A H, then the dotted line A H represents the resultant of all these forces. We shall now show how we can equally well apply this to E.M.F.'s acting in a circuit. In an alternating-current circuit we have usually three E.M.F.'s acting :

(1) The effective E.M.F., or the product C R.

(2) The E.M.F. due to self-induction or inductance.

(3) The impressed E.M.F., which at any instant balances the other two.

Suppose we have a line, o A (in Fig. 41), revolving about the point o ; if o A represents to some scale the maximum value of an E.M.F. undergoing a sine variation, it will be seen that we may represent the values at any instant by the

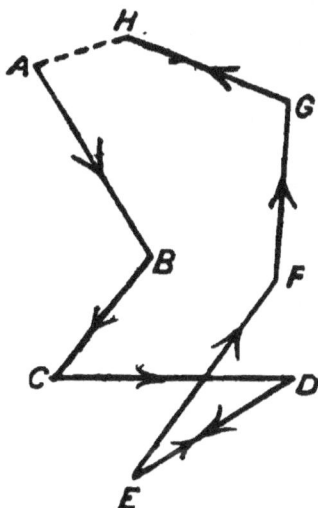

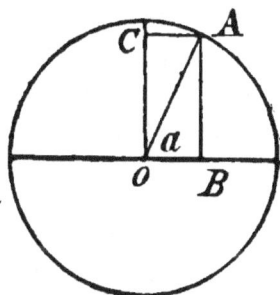

FIG. 40.   FIG. 41.

projection of the line o A upon a fixed line, such as o C, which it will be seen is proportional to sin a. Now o A may represent the resultant of any number of E.M.F.'s. We may extend this, as in the case of the force polygon, and state that the projection of the resultant of any number of E.M.F.'s is the sum of the projections of its constituents.

An inductive alternating-current circuit is a much more complicated matter to deal with than the other, since we have to consider the disturbing effects of self-induction to obtain the power consumed in an inductive circuit. We must not multiply the $\sqrt{\text{mean square}}$ of the alternating current by the $\sqrt{\text{mean square}}$ of the alternating E.M.F., because the product would not measure the true mean power. The mean power at any instant of time is obtained by multiplying the current at that instant by the E.M.F. at that instant; the same must be done for any and every other instant. We thus have a series of simultaneous results, and the mean value of this series will give the true mean power consumed. The product of $\sqrt{\text{mean square}}$ current and $\sqrt{\text{mean square}}$ E.M.F. is called the *apparent* mean power consumed in an inductive circuit; whilst the mean of the products of current and E.M.F. is called the *true* mean power consumed in an inductive circuit.

For a non-inductive circuit, or when the current curve is in step with the E.M.F. curve, the $\sqrt{\text{mean square}}$ current × $\sqrt{\text{mean square}}$ E.M.F. = mean product of current and E.M.F.; but, as explained above, this is not so with an inductive circuit, or when the current curve lags behind the impressed E.M.F. curve. In some cases the true watts may be only one-half what the apparent watts are.

In an inductive circuit we have to deal with self-induction, which may be defined as the E.M.F. that is produced by the varying current acting inductively on its own circuit. Where the current is a steady one—that is, of constant value—there is no effect apparent of self-induction, because the magnetic field enclosing the current is invariable in strength; but immediately the current begins to change and vary in value, like an alternating current, its magnetic field also begins to change in value, and when a conductor is surrounded by a magnetic field which is constantly varying in strength, an E.M.F. is set up which is opposite in direction to the impressed E.M.F. of the circuit; thus, in addition to the mere resistance in ohms of the conductor, there is a second obstruction to be dealt with. The obstructive effect which this counter E.M.F. has upon the current of the circuit is named the "inductance" of the circuit. Hence, we may say that a circuit has unit inductance when a current varying at the rate of one ampere per second will produce in it an E.M.F. of one volt, and this unit is named a "henry," or a "secohm," and is generally denoted by the letter L.

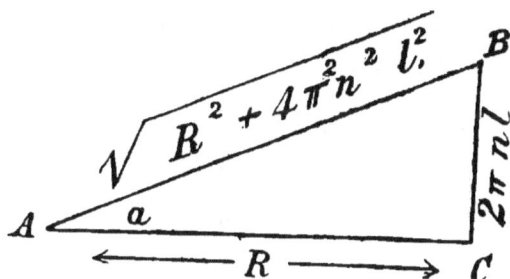

FIG. 42.

When the periodicity of the current is $n \curvearrowleft$ per second, then the inductance is measured by $2 \pi n L$; where L is the said coefficient of self-induction, let the resistance of the conductor be R ohms, then the total obstruction to the current is $\sqrt{R^2 + (2 \pi n L)^2}$, and this is named the impedance of the circuit, since it represents the whole resistance which the current has to contend against. Fig. 42 shows the relation between inductance and resistance, where resistance and inductance occupy two sides of a triangle, $A C = R$ and $B C = 2 \pi n L$, then the hypothenuse, or third side, A C, must equal the $\sqrt{\text{sum of the squares}} = \sqrt{R^2 + 4 \pi^2 n^2 L^2}$; hence the side A B = the impedance of the circuit. Suppose in a circuit having inductance whose coefficient is L, resistance = R, and impressed E.M.F. = E, that the maximum value of the current be I, then the product R I represents the volts-absorbed by the metal resistance of the circuit, and these volts are named the effective E.M.F. The product $(2\,\pi\,n\,L)$ I will represent the E.M.F. due to the inductance of the circuit, whilst the product $\sqrt{(R^2+4\,\pi^2\,n^2\,L^2)}$ I will represent the impressed E.M.F. of the circuit. The angle B A C in Fig. 42 always defines the lag of the current I behind the impressed E.M.F.—that is, it measures the angle of retardation in phase — while its tangent represents the ratio existing between the inductance and resistance of the circuit. We can now give a rough formula for the adaptation of Ohm's law for alternating currents, which is—

$$\text{Current} = \frac{\text{Impressed E.M.F.}}{\text{Impedance}}.$$

Mathematically this may be expressed—

$$C = \frac{E \sin \theta}{\sqrt{R^2 + 4\,\pi^2\,n^2\,L^2}}.$$

The following experiments have been made by Mr. W. Brew, of the British Museum, and have been kindly supplied to the writer as illustrating some interesting results on the skin effect of alternating currents in conductors.

Skin Effect.—The method of experiment consisted in observing the increase in the force exerted upon a magnet pole reversing in synchronism with the alternate current, and placed just exterior to the conductor in which it was flowing. As the rate of alternation of the current was increased the current was forced outwards to the external layers of the conductor, and the force exerted upon the magnet pole was correspondingly increased.

In one experiment, using a solid brass conductor 1in. in diameter, and passing various currents of a frequency of 200 through it, it was found that the force on the magnet pole 'had increased by about 20 per cent. of its original amount with continuous currents of equal value flcwing in the conductor.

In another experiment a brass tube of the same external diameter, but ¾in. internal diameter, was employed, and some of the results obtained are plotted in the accompanying curves, shown by Fig. 43, having for abscissæ the frequency of the alternating current and as ordinates the percentage increase in the force exerted upon the magnet pole over that due to a continuous current of the same numerical value.

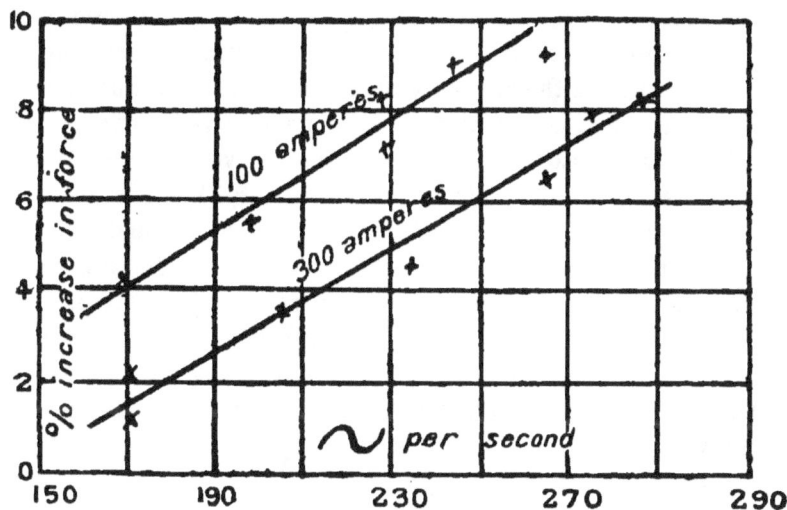

FIG. 43.

The upper curve was obtained with a current of 100 amperes at various frequencies, and the lower curve refers to a current of 300 amperes passing through the conductor under similar conditions.

The curves show us, amongst other things, that with a low-current density the skin effect becomes more apparent than with a high one; also that after about 16,000 alternations per minute, or a frequency of 266 per second has been reached, the current rapidly becomes forced to the external layers of the conductor.

Transformers.

From what has been stated regarding the features of high and low pressure distribution, it is at once apparent that the best way to distribute electric power over large areas with a limited loss, and also at the same time to introduce it into the consumer's premises at a low and safe voltage, is by using a combination of pressures, a high pressure being used for "distributing" purposes, and a low pressure being used on the "lamp" circuits. This can be effected in alternate-current work by the aid of what are named "transformers," and their object is to transform or convert small currents of high voltage into large currents of low voltage, and *vice versâ* ; so that the electricity is sent out into the distributing mains at a high pressure, and where a tapping is made for a consumer, a transformer is connected up to the primary circuit to receive the high-pressure current while it gives out a low-pressure current, which is led into the secondary circuit, the current being thus transformed. A transformer which changes a current from a low pressure into a high pressure is termed a "step-up" transformer, and a transformer which changes a current from a high pressure into a low pressure is termed a "step-down" transformer.

Except for a small loss, the power received and the power given out are the same, so that when the pressure is decreased the current is increased ; for example, a high-pressure current of 20 amperes at 2,000 volts is led into a "step-down" transformer, and if the winding on the core is such that the pressure is decreased to 200 volts then the current will be increased to 200 amperes, because $2,000 \times 20 = 200 \times 200$ = watts. In its simplest form a transformer takes the form of an annular ring of iron, having a great number of turns of thin wire wound on one side and a small number of turns of coarse or thick wire wound on the opposite side—the thin wire is named the "primary" coil, and the thick wire the "secondary." The high-pressure current being led through the primary coil induces a low-pressure current in the secondary, or a low-pressure current being led through the secondary will induce a high-pressure current in the primary. These two windings are entirely distinct from each other, and well insulated from one another. Instead of the two windings being on opposite sides of the ring, one can be wound on top of the other, in which case it is usual to wind the primary on first, the secondary being wound on top of the primary. Fig. 44 shows a ring with the primary and secondary on opposite sides, A representing the primary and B the secondary.

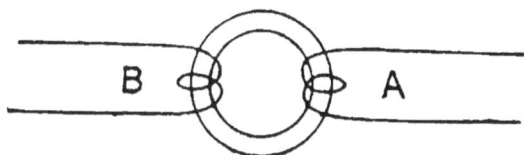

B A

FIG. 44.

It is thus seen that an alternate-current transformer consists simply of copper and iron—copper to carry the electric current, and iron to carry the magnetic flux. The ratio between copper and iron for transformer design is a very nice point. As yet there does not seem to be sufficient practical data to warrant any decided ratio, and, in addition, this ratio would not be the same for all kinds of transformers : it would vary with the nature of the design, and also with the require-ments of the transformer. Some design for high efficiency at heavy loads, others strive to produce one that will possess a high efficiency for low loads ; others, again, prefer to design one that will give the best average efficiency for all loads.

W

To give an idea, it may be assumed that a fair ratio between
copper and iron is 1 : 2, and that 1lb. of copper and 2lb. of
iron, or total of 3lb., will yield an output of 60 watts, or
sufficient energy for one incandescent lamp of 16 c.p. This
is at the rate of 20 watts per pound dead-weight, and applicable to a medium-sized transformer; one that would run,
say, 100 incandescent lamps of 16 c.p., and thus have a total
weight of 300lb.

It is necessary to fix transformers in a fireproof place,
since they heat up very much when at work, the iron core
becoming hot through "hysteresis" effects of the alternating
current. This heating signifies so much loss of energy, and
the greater this loss the less is the efficiency of the machine.
The best modern transformers will now give an "efficiency"
of between 96 and 97 per cent. at full load, and 92 per cent.
at one quarter load. These figures are very satisfactory, and
there does not seem to be much room for improvement as
regards this efficiency at full load and half load. It is only
when they are worked at light load that they become terribly
wasteful, unless they are designed especially for light loads,
and in that case they are not efficient with heavy loads.
What is wanted is a transformer that will give a good
efficiency for all loads, from full load down to a few lamps.
The iron core of the transformer is alway laminated, or built
up, by using a number of thin sheets, as is the case in the
construction of all alternate-current apparatus that contains
iron, each sheet being insulated from the other by shellac,
paper, etc., or some such suitable insulating material.

In high-tension alternating-current systems of distribution
it is the present practice to "bank" transformers. By
"banking" is meant grouping together in sub-stations.
Thus there are two networks of distributing mains—(1) the
high-pressure primary mains that ramify from the generating
source, and (2) the low-pressure secondary mains that branch

out from several determined centres, positioned at suitable places throughout the area to be supplied; it is at these centres where the sub-stations are built, each having a number of transformers, some large and some small. As the load increases, so transformers are switched into circuit, and as the load decreases they are switched out; hence it is necessary to have an attendant at each sub-station to watch the load and perform the above operations. By this device, no transformer need be at work with a light load, and by the aid of one or two small ones, to put in circuit when the load will not justify the employment of a large one, the losses are reduced considerably. It may be remarked that the core losses, or the loss of energy due to the heating of the iron, is almost independent of the loads, and so when the load is light, they form much the greater part of the total loss of energy in the transformer. To protect the primary and secondary windings from an undue current, suitable fuses are inserted in the terminal-box of the transformer, a fine fuse being used for the primary and a thick one for the secondary. It is usual also, in some cases, to have the secondary coils wound in two windings, so that by the aid of a connecting strip in the terminal-box, the secondary can be at once connected up either in parallel or in series. In the former case the voltage may be, say, 50 volts, and in the latter case it would then be 100 volts, the transformer being applicable to either of the two pressures. It is not often in practice that the primary fuse "blows"; if it does, care should be taken that the primary coil be cut out of the high-pressure circuit before an attempt is made to replace the fuse. The secondary fuse is much more liable to blow, as it is direct on the lamp circuit.

Sub-stations are expensive places to build and equip with apparatus, and, in addition, the necessity of having attendants in charge makes their proper maintenance still more expensive;

w 2

so that the presence of the transformer in alternate-current distribution materially reduces the saving that is gained by the employment of small conductors. Some go so far as to say that the cost of transformers comes to more than the saving that is effected in the cost of copper by employing small conductors for high-pressure currents, but however true this may be for limited areas, there seems to be no mistake about the fact that for large and extensive areas high-pressure alternating currents are preferable to three or five wire direct currents, in order to obtain a distribution of power in an economical manner and, above all, with effective and good regulation for constant lamp pressure. The following figures relate to the construction of Labour's transformer, a type which may be taken as a good representative machine designed for high efficiency at all loads. The output is for 15,000 watts, and so it is capable of supplying current to about 250 lamps of 16 c.p. The primary coils are wound for a primary current of from 2,400 to 2,500 volts, and the secondary are wound for a current of about 100 to 105 volts.

Primary Circuit.

Cross-section	4·52 sq. mm.
Current, about	6·25 amperes.
Current density	1·38 per sq. mm.
Number of turns	720
Resistance	2 ohms.

Loss of E.M.F. on circuit = 12·5 volts = ·5 per cent.

Secondary Circuit.

Cross-section (stranded)	133 sq. mm.
Current, about	150 amperes.
Current density	1·13 per sq. mm.
Resistance	·00333 ohm.

Loss of E.M.F. on circuit = ·0499 volt = ·5 per cent.

Weight of machine .... 495lb.
Cross-section of iron .... 150 sq. cm.
Primary E.M.F. .... 2,400 volts.
Frequency per second 80
Induction per sq. cm. 4,440
Maximum induction 6,250

The variation of E.M.F. between open circuit and at full load is about 1 per cent. when the primary E.M.F. is constant.

The efficiency calculated for a mean temperature is—

at $\frac{1}{8}$ of full load .... = 86 per cent
at $\frac{1}{5}$ „ .... = 90 ,
at $\frac{1}{4}$ „ .... = 92 ,
at $\frac{1}{3}$ „ .... = 94 „
at $\frac{1}{2}$ „ .... = 96 „
Full load .... = 97 „

The waste of power on open circuit is less than 2 per cent. of power yielded when working at full load.

Insulation.

The object of insulating matter put on copper conductors is to confine the current as much as possible to the copper, and to allow as little as possible to flow through other paths and so leak away. The leakage of water in a pipe is somewhat analogous to the leakage of the electric current flowing along a conductor : the longer the pipe and the more joints there are in it, the greater will be the chance of leakage ; and it is similar with electricity. The two typical substances for conducting and insulating purposes as used in practice are copper and indiarubber. The conductor is made of copper, because it is desirable to have the resistance to the current as low as possible ; the insulation is made of indiarubber, because it is desirable to obstruct the current as much as possible. The copper conductor is made of such a sectional area that a certain current can flow without unduly heating the metal; and since the specific resistance of indiarubber is enormously higher than that of copper, it follows that only a most minute current can leak through the insulation. The conditions under which electricity is supplied reduce considerably the great difference between the two, and instead

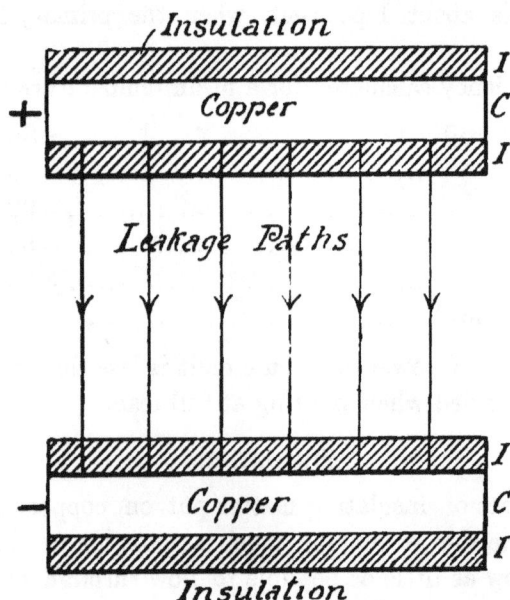

FIG. 45.

of the ratio of the specific resistances of indiarubber and copper being billions to one, it is brought down to a million to one, which is a ratio we are more capable of conceiving.

We will consider the case of a couple of conductors, say, 200 yards long, and take the ratio of the specific resistances of insulation and copper as $6 \times 10^{16} : 1$. Let the depth of the insulation be, say, ·5cm., and the diameter of the copper 1cm., as shown in Fig. 45. Considering the insulating matter first, the length of its path is extremely short, while the cross-sectional area is enormous. The reason of this is because the current does not tend to flow along the cylindrical tube of insulation, but tries to flow *through* the insulation covering *in the direction of its depth*, and thus through the air space or whatever intervenes between the two conductors, then through the depth of the insulation of the second conductor, and so complete its circuit. This leakage does not happen at one point or many, but at every point along the conductors, and at every point around the circumference. This being so, the sectional area of the insulation is measured by the cylindrical surface exposed by the whole of the conductors, and this surface is obtained by multiplying the length of the conductor by the mean circumference.

The total length of the insulating path is extremely short, being evidently the depth of the insulating matter that covers the copper conductors, and as there are two conductors so this total length is measured by twice the depth of the insulation. The intervening space between the copper conductors cannot always be calculated as part of the insulating path, because the insulation on each conductor is in contact with earth at some points. And even when the conductors are suspended in air, or embedded in bitumen-filled pipes, there is bound to be some supporting points where contact of the insulation with earth occurs.

Let l signify length of cable;

    D ,,   outer diameter of insulation;

    d ,,   inner   ,,   ,,   ,,

Then $\frac{1}{2}$ (D – d) signifies thickness of insulation, and d signifies diameter of conductor.

The sectional area of the conductor $= \pi\, d^2 \div 4$;

  ,,   ,,   ,,   ,,   insulation $= \left(\dfrac{D+d}{2}\right) \pi\, l$;

The total length of the two conductors $= 2\,l$;

,,   ,,   ,,   ,,   insulation $= (D - d)$.

We will assume, to simplify matters, that the two cables are buried, say, a foot apart in the ground, which we will further assume is slightly moist, so that the outside surfaces of the insulating matter on each cable is "to earth" in a most evident manner. The resistance of the path of the leakage current is measured simply by the thickness of the insulation on each cable, because the resistance offered by the "earth" is of a negligible quantity; the thickness of the insulation being $\frac{1}{2}\,(D - d)$, and there being two thicknesses to penetrate—one on the positive cable and one on the negative—the total path $=$ twice the thickness, or $(D - d)$. The mean diameter of the tubular insulating covering is evidently $\frac{1}{2}\,(D + d)$; hence the mean circumference is obtained by $\frac{1}{2}\,(D + d)\,\pi$, and multiplying this by the length, l, gives the mean sectional area $= \frac{1}{2}\,(D + d)\,\pi\,l$. Inserting the respective values, we thus have :

For conductor $R_c = \dfrac{8\,l}{\pi\,d^2} \times \rho$;

For insulation $R_i = \dfrac{D - d}{l\,\pi\left(\dfrac{D + d}{2}\right)} \times \rho_1$;

where ρ and ρ_1 signify the relative resistances of the copper and insulation covering respectively.

It is impossible to say what the resistance of insulating covering might be. The maximum would be if the covering were made entirely, say, of indiarubber, but of course it never is—the number of layers of indiarubber put on varying with the quality and make of the cable. Copper being taken as unity, we will assume, therefore, that relative resistance of the covering, is, say, 6×10^{16}; that of indiarubber being about 6×10^{20}.

Hence, the relative resistances of the insulation and copper are $6 \times 10^{16} : 1$. Inserting the value of 6×10^{16}, we have the ratio between insulation and conductor:

$$6 \times 10^{16} \left\{ \frac{D-d}{l \, \pi \left(\dfrac{D+d}{2} \right)} \right\} : \frac{8 \, l}{\pi \, d^2} \; ;$$

or, $\qquad 6 \times 10^{16} \left(\dfrac{D-d}{D+d} \right) : \dfrac{4 \, l^2}{d^2} \; ;$

or, $\qquad 1 \cdot 5 \times 10^{16} \left\{ \dfrac{d^2}{l^2} \, \dfrac{(D-d)}{(D+d)} \right\} : 1.$

Fig. 45 gives a sectional view of a pair of insulated conductors showing the leakage paths across from copper of one conductor to the copper of the other. From this it can readily be understood how it is that when mains laid along casing which becomes moistened or damp the leakage is increased. Although the difference between the specific resistance of the insulation and that of the conductor is so enormous, yet it is reduced quickly. The total length of the conducting path was assumed to be 400 yards, and the length of the insulating path is taken as twice the depth of insulation, or $2 \times \cdot 5 = 1\,\mathrm{cm}.$, so that the latter path is only about

$$\frac{1}{400 \times 36 \times 2 \cdot 54} = \frac{1}{36,576} \text{ of the length of the former.}$$

The ratio at the start was taken as $6 \times 10^{16} : 1$, and this now becomes reduced at once to 36,000 times smaller, because the conducting path is about 36,000 times as great as the insulating path. Again, we must consider the ratio of the sectional areas of the two paths. The sectional area of conducting path is $\dfrac{3 \cdot 1416 \times 1^2}{4} = \cdot 7854$ square centimetre.

The sectional area of the insulating path is the mean circumference of insulation × length of conductor = 1·5 × 3·1416 × 200 × 36 × 2·54 = 86,180 square centimetres. Thus the sectional area of the second is 110,000 times as great as that of the first, and consequently the resistance of the insulating path will be reduced nearly 110,000 times, and the ratio is now further reduced, so that finally the insulating path may be taken as having a resistance about 1·5 × 10⁷ times as great as that of the conducting path, and hence a most minute part of the total current leaks away through the insulation, and is wasted. All these figures are very approximate to the case considered, and are given simply to give an idea of the relation that may exist between the insulating and the conducting paths. The amount of the leakage current is about the best test we can have respecting the state of the insulation on an electric light circuit.

It is of the utmost importance that every part of the circuit of an installation should be well insulated, otherwise there will be continual trouble—the current will leak, fuses get blown, and the short-circuits may overheat the wires and cause a fire to break out. The state of the atmosphere has a great deal to do with the condition of the insulation of the circuit. If the weather be fine and dry, very little leakage will be discovered on testing ; but test the same circuit on a wet day, when the atmosphere is charged with moisture, and very different results are sometimes obtained. The damp air produces moisture on the surfaces, and so enables the current to creep or leak. This "surface leakage" is not confined to such material as wood, but also acts on such insulators as glass and porcelain, materials which in dry weather form excellent insulators. It must be remembered that England is a rainy country, and has more often a moist than a dry atmosphere—in fact, we rarely get a dry atmosphere such as is known in warmer climes, so insulation which will do for other countries will not do for us. In addition to the leakage itself, moisture tends to set up electro-chemical action, and the insulation and the copper thus get acted on chemically and destroyed. All circuits should be systematically and frequently tested, so that the slightest sign of any incipient fault becoming developed can be detected as soon as possible and the fault removed before it can do any harm. The insulation resistance of electric light cables is quoted as so many megohms per mile. The word *per* must here be interpreted in its true meaning, which is "divided by," because insulation resistance varies inversely as the length of the cable— that is, if one mile of cable has a resistance of 1,000 megohms, then half a mile will have an insulation resistance of 2,000 megohms, and two miles 500 megohms, etc. ; hence the insulation resistance should be expressed as one factor of a product. Thus 1,000 megohm-miles means one mile will be 1,000 megohms, or 1,000 miles will be $\frac{1}{1000}$ of a megohm = 1,000 ohms.

High-insulation cables are made to give 1,000 megohm-miles, and are insulated with vulcanised indiarubber, which material seems to find most favour for insulating purposes. A cable of this class would be suitable for high-pressure currents. For low-pressure currents the insulation resistance need not be so high, 300 megohm-miles being reckoned sufficient.

Insulated cables have now been brought to a high standard of perfection, particularly since high-pressure currents have been in general use, and where the conductor is required for damp places the exigencies of the case have always been successfully coped with. Each manufacturer has his own ideas respecting the best method of building up the insulation covering, and we think we cannot do better than to append the valuable tabulation drawn up by Mr. W. H. Preece as given in his paper, entitled "Specification of Insu-

lated Conductors for Electric Lighting and other Purposes,"
read before the Institution of Electrical Engineers on
December 10, 1891. See Tabulation 35.

There seems to be no definite rule respecting the state of
insulation resistance for electric light circuits, which is to be
regretted.   Each electric supply company requires a different
standard, the same with the insurance companies ; and there
is a great variety in the way of expressing it, some requiring
an insulation resistance a dozen times more than others, so
that these rules are to a great degree very arbitrary.   When
treating of insulation resistance, it must be remembered that
the greater part of the leakage is caused by the " fittings "—
such as fuses, switches, lampholders, etc. ; these harbour
moisture, and so promote surface leakage : this is why it
makes such a difference whether the insulation is tested on a
dry or a wet day.   The most general practical formula for
insulation resistance is of the form—

Rule (I.) $R_i = K \dfrac{E}{C},$

where K is some constant the value of which differs according
to different authorities ; and whatever this value of K may
be, the result is that the insulation resistance must, according
to the above formula, be some function of the resistance of
the conducting circuit.

With a fixed potential difference, E, the resistance of the
circuit determines the useful current in the conductors, while
the insulation resistance determines the leakage current ; and
as the insulation resistance is K times the circuit resistance,
hence $\dfrac{1}{K}$ measures the leakage current as the fractional part
of the useful current.   For example, let the pressure of the
circuit be 110 volts and the current, say, 275 amperes,
supplying about 500 lamps of 16 c.p. ; then, if a numerical

TABULATION 35.—SPECIFIC INSULATION (σ) OF ELECTRIC LIGHT CABLES MANUFACTURED BY VARIOUS FIRMS.

Manufacturer.	No. of Wires and approximate gauge.	Conductor. Diameter. inch.	Conductor. Diameter. mm.	Dielectric. Thickness (t). inch.	Dielectric. Thickness (t). mm.	Dielectric. Diameter (D). inch.	Dielectric. Diameter (D). mm.	Insulation per mile at 60 deg. F., megohms.	Specific Insulation (σ).	Remarks on Insulation.
Siemens Bros.	7/22	·084	2·14	·076	1·93	·236	5·99	16,180	15·84	Siemens high-insulating Indiarubber.
,,	19/16½	·340		·078	1·98	·240	6·10	18,270	17·60	Ordinary pure and vulcanised rubber.
,,	61/15½	·612	8·64	·079	2·03	·501	12·72	1,022	2·67	
,,	61/12	·850	15·54	·168	4·27	·949	24·10	824	1·90	
,,	37/13	·651	16·53	·190	4·82	1·230	31·24	4,082	11·17	Lead cased; high-insulating fibrous material.
,,	19/14	·410	10·41	·086	2·44	·844	21·43	2,507	10·00	
,,	7/13	·286	7·26	·079	2·01	·568	14·42	3,734	11·77	
,,	60/6½	1·656	42·06	·118	2·87	1·882	47·80	6,389	14·69	Lead cased; impregnated fibrous material.
,,	60/10	1·170		·114	2·89	1·398	35·51	624	4·93	
Silvertown	19/15	·360	9·14	·144	3·66	·648	16·46	830	6·56	Pure indiarubber, then vulcanised indiarubber.
,,	,,							851	4·83	
,,	,,							1,014	5·76	
Callender	7/16	·192	4·88	·154	3·91	·500	12·70	6,768	11·64	Vulcanised bitumen.
,,	19/18	·240	6·10	·155	3·94	·560	13·97	6,664	11·46	
Glover	19/13	·460	11·08	·115	2·92	·690	17·52	6,549	11·26	Pure and vulcanised rubber.
,,	,,							380	·40	
,,	,,							400	·49	
,,	7/16	·192	4·88	·055	1·40	·302	7·67	600	1·50	
,,	,,							515	1·28	
,,	,,							560	1·40	
Fowler-Waring	1/18	·048	1·22	·036	·91	·120	3·05	855	1·91	Waring compound.
,,	1/16	·064	1·62	·043	1·09	·160	3·81	813	1·81	
,,	7/16	·192	4·88	·078	2·03	·352	8·94	833	1·86	
,,	19/18	·240	6·10			·400	10·16	6,522	7·20	
,,	19/18	·320	8·13	·096	2·41	·610	12·95	6,619	7·85	
,,	19/15	·360	9·14	·100	2·64	·560	14·22	4,835	8·07	
Henley	1/10	·128	3·25	·046	1·17	·220	5·59	3,280	6·48	Pure and compound indiarubber.
,,	7/16	·192	4·88	·054	1·37	·300	7·62	3,312	7·18	
,,	19/18	·240	6·09	·060	1·52	·360	9·14	3,145	7·19	
,,	1/14	·080	2·03	·040	1·02	·160	4·06	564	1·05	
,,	3/30	·078	1·98	·039	·99	·157	3·99	406	·92	
,,	7/15	·218	5·49	·057	1·45	·330	8·38	329	·82	
,,	19/14	·400	10·16	·080	2·03	·560	14·22	4,440	6·48	Ozokerited Indiarubber.
,,	61/15	·648	16·47	·111	2·82	·870	22·09	5,210	7·57	
,,	61/12	·936	23·77	·217	5·51	1·370	34·79	2,210	5·27	
,,	,,							2,000	6·01	
,,	,,							7,146	19·98	Special cable—Class AA.

value of 10,000 be given to the constant K, we shall have

$$R_i = \frac{10,000 \times 110}{275} = 4,000 \text{ ohms.}$$

Hence R_i, the insulation resistance, works out to 4,000 ohms, and as K has a value of 10,000, therefore the leakage current will be $\frac{1}{K} = \frac{1}{10000}$ part of the lamp current, or in this case $\frac{275}{10000}$, or, roughly, $\frac{1}{36}$ of an ampere.

The above-mentioned formula is probably the simplest and most commendable to use, because it defines at once the ratio between the leakage current and the total current in practical work. If this is known, it tells us pretty well the condition of the insulation. With regard to the value to be given to K, the value ranges from 5,000 to 50,000, or even 100,000.

Instead of judging of the insulation resistance by its leakage current, it is a good rule to specify so many megohms as a constant and then to divide by the number of 16-c.p. lamps there are on the circuit. The rule then being

Rule (II.) $R_i = K \div$ lamps.

The constant, K, in use amongst electric supply companies for 100-volt circuits varies from 100 to 10 megohms. For example, take a circuit of 100 volts, having, say, 1,000 lamps of 16 c.p. in parallel, and let the constant be 50 megohms, then the insulation resistance will be

$$R_i = 50,000,000 \div 1,000 = 50,000 \text{ ohms.}$$

Upon working out, this resistance will be found to correspond with a leakage current of $\frac{1}{300000}$ part of the total current.

Several of the leading electric supply companies adopt this rule, giving various values to K. Thus, the City of London

require the high value of 75 megohms, the Bradford Corporation are satisfied with 10 megohms, while the Glasgow
Corporation demand 60 megohms. Amongst those who
prefer the first rule, that of stating the leakage current as a
fractional part of the useful current, are several insurance
companies, who have settled on $\frac{1}{20000}$ part of the current,
while the theatre restrictions made by the London County
Council allow a leakage of $\frac{1}{15000}$ part.

It is customary to double the constant when alternating
currents are used, because the maximum value of the E.M.F.
is higher than the working voltage, which, as before mentioned,
is expressed by $\sqrt{\text{mean square}}$.

A third rule is a mixture of the two previous ones, and
consists in multiplying the constant K by the working
voltage, whatever this may happen to be, and dividing by
the number of lamps of 16 c.p. that are wired. This rule
thus takes the form—

Rule (III.) $\qquad R_i = K \dfrac{E}{N}.$

Take a plant running, say, 4,000 lamps at 110 volts
pressure, putting K as 80,000, we obtain

$$R_i = 80,000 \times \frac{110}{4,000} = 2,200 \text{ ohms.}$$

Tabulation 36 gives the insulation resistance worked out
from Rule (I.) for various currents from 10 to 2,000, and
also at various voltages, the standard for K being put at
50,000, so that the figures given are obtained from the
formula $R_i = 50,000 \dfrac{E}{C}$, and thus show what the insulation
resistance in each case should be, so that the leakage current
will not exceed $\frac{1}{50000}$ part of the total current. This

standard may be taken as one that will give most satisfactory
insulation.

TABULATION 36.

Maximum Current.	Working Pressure of Circuit in Volts			
	110	80	65	50
10	550,000	400,000	325,000	250,000
30	183,330	133,330	108,330	83,300
60	91,700	65,000	54,166	41,660
100	55,000	40,000	32,500	25,000
200	27,500	20,000	16,250	12,500
400	13,750	10,000	8,125	6,250
600	9,100	6,500	5,416	4,160
800	7,000	4,000	4,621	3,120
1200	4,600	3,330	2,708	2,080
1600	3,400	2,500	2,310	1,560
2000	2,750	2,000	1,625	1,250

Testing Circuits.

When the leakage current exceeds the fixed amount
allowed, or when the insulation resistance falls below the
fixed standard, whatever this amount or standard may be,
then the circuit must be considered as faulty or defective,
and should at once be tested so as to locate and remove the
fault. Sometimes the falling away of the insulation resistance cannot be traced to any particular point. In that case,
it is most probably a general deterioration of insulation
throughout the circuit, and nothing can be done except take
down what appears to be the worst section, and replace with
new mains. Faults may be divided into two classes—(I.)
partial earths, (II.) dead earths.

A partial earth signifies some weak spot in the insulation
covering where the covering is partially worn away, due to
corrosive action, abrasion, or mechanical damage of any kind.
A partial earth may also be caused by "wet rot." This condition of the insulation is usually brought about by damp
state of the atmosphere or walls, or dripping water, action
of steam, etc. The insulation then becomes softened and
"soddened," and when the two cables get touching, or close

together, the leakage current becomes abnormally great; and with wet casing, electric sparks will flash intermittently across the path. This leakage and minute "arcing" will eventually overheat the insulation and casing, which will probably burst into flame.

A dead earth is obtained when there is actual metallic connection between either of the mains and "earth"; by "earth" is signified any system of water or gas pipes, or any such unconfined conducting or semi-conducting medium that would allow the current to leak away; it also means the actual soil. Of course, when there is a dead earth on both mains, a short-circuit is produced, for the current will enter earth by one main and leave by the other.

A rough test for earths is shown in Diagram I. of Fig. 46. G and B signify a linesman's galvanometer and battery, one pole or wire of which is put to earth, and the other is joined on to one of the mains, which is disconnected from the other main by the removal of all lamps; then, if there should happen to be an earth on the main to which the galvano-meter is connected, the needle will show a greater or less deflection, because the circuit will be completed. In a similar way, an earth or fault can be detected in the other main by joining up the galvanometer to that main. This simple test does not give any quantitative information respecting the fault: it only shows in a rough way that a fault is there.

To obtain a quantitative test, or measure the insulation resistance of the circuit, some such testing set like the "Silvertown" one must be used. This set is, perhaps, the best there is for rough practical use, such as in central-station work, and it is particularly handy for this sort of work. It is described and illustrated in several books on electri appliances, and those who do not know it can find it in Jamieson and Munro's pocket-book. The method of using

x

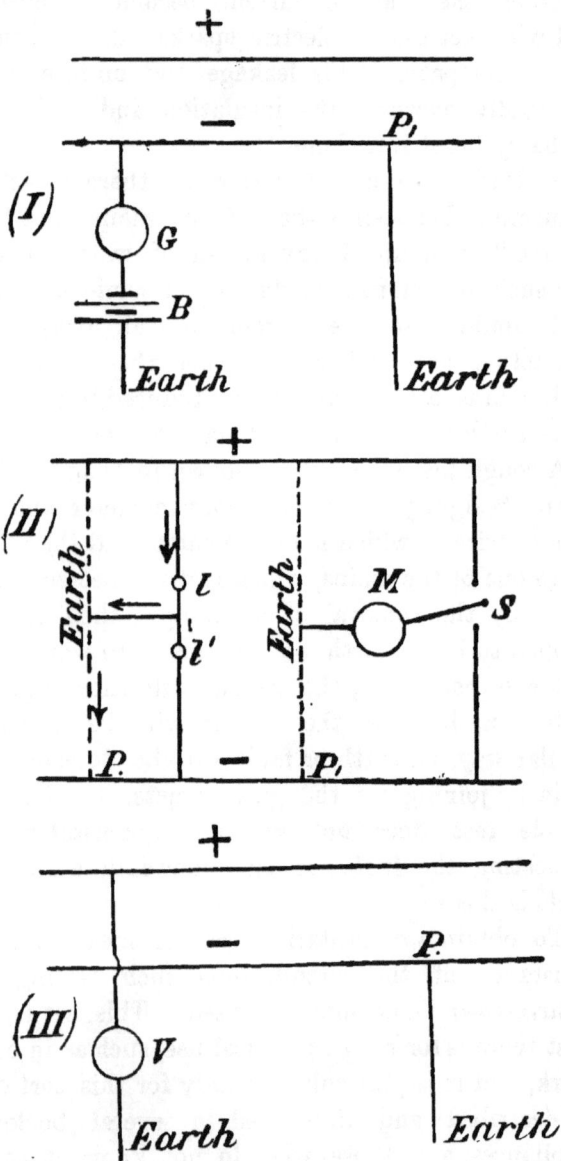

Fig. 46.

this set is as follows : Insert the two battery wires into the two terminal holes marked "insulation," and having first adjusted the needle of the galvanometer to zero by means of the controlling magnet, then plug up the hole marked "10,000 ohms," and observe what deflection is given and what shunt is used. The best combination of battery power and shunts to be used must be guided by the probable value of the insulation resistance, because to obtain the best results the deflection obtained from the insulation resistance must be made to coincide as near as possible with that obtained from the 10,000 ohms. In addition, it is always best to obtain a deflection as near as possible to 45deg. When this operation has been done, one of the mains of the circuit about to be tested is connected up to the screw terminal marked "insulation," placed on the left hand side of the galvanometer. A wire is then attached to the right-hand screw terminal, marked "earth," and its other end is fixed on to a gas or water pipe, to make a good earth. To get a bright and clean contact, always scrape away the surface or skin of the pipe, as the test depends on having a good "earth." All that now remains to be done is to connect up the battery wires as before and plug up the hole marked "insulation," which is next to the 10,000-ohm hole. Note the new deflection and the shunt, if any shunt be used.

There are thus three variables in use : first, the number of cells ; second, the shunt used ; third, the value of the deflections.

With cells and shunt constant, $R_i \propto$ inversely as deflection.
With cells and deflection ,, $R_i \propto$,, ,, shunt.
With shunt and deflection ,, $R_i \propto$ directly as cells.

The above statements can easily be reasoned out, R_i signifying insulation resistance as compared to the standard resistance of 10,000 ohms. These can be expressed in the following formula :

$$R_i = \frac{D_1}{D_2} \times \frac{S_1}{S_2} \times \frac{N_2}{N_1} \times 10,000,$$

where D_1, S_1, and N_1, refer to the deflection, shunt, and number of cells when using the 10,000-ohm plug, and D_2, S_2, and N_2, the same when using the insulation-resistance plug.

When using the 10,000 ohms, it is usual to obtain what is termed a constant. This is obtained by multiplying the 10,000 ohms by the deflection, then by the value of the shunt, and then dividing by the number of cells used. Putting K for constant, this gives us

$$K = \frac{10,000 \times S_1 \times D_1}{N_1}.$$

When this constant is found, the formula for the insulation resistance becomes

$$R_i = \frac{K\ N_2}{D_2\ S_2}.$$

The two shunt plugs are marked $\frac{1}{9}$ and $\frac{1}{99}$; this signifies that when the $\frac{1}{9}$ plug is inserted, the multiplying power of the shunt is 10—that is to say, that only one-tenth of the total current passes through the galvanometer, the remaining nine-tenths passing by. Similarly, when the $\frac{1}{99}$ plug is in, the multiplying power of the shunt is 100.

Diagram II. of Fig. 46 illustrates a very good way of obtaining a continual indication respecting the state of the insulation resistance. Two lamps of 16 c.p., l and l', are connected in series across the mains, and the wire connecting them in series is lead to earth. The brilliancy of the lamps indicate the state of the insulation resistance of the circuit. In central stations the lamps can be connected to the two omnibus bars where a low-pressure parallel system is used. Suppose some part of the − conductor makes an earth somewhere, as shown at P, then a flow of current will take place from the + conductor through lamp l to earth, and thence to the − conductor, as shown by the arrows ; consequently, the lamp will glow more or less brilliantly according to the quantity of current flowing through it, and this will depend upon the resistance there is between the earth and the − conductor. The same thing occurs when the + conductor has an earth on it ; then, however, the lamp l' will glow instead of the lamp l, so that the brilliancy of that lamp connected to one conductor will indicate the state of the insulation of the other conductor.

When both lamps glow equally, then the insulation resistance of each conductor is equal. The lamp, however, only gives a rough test, and in order to obtain some measurement of the leakage current, it is necessary to employ a low-reading ammeter. This method is shown in Diagram II. also ; for in place of the two lamps the ammeter, M, is placed, having a switch, S, so that either the + or the − side of the circuit can be connected up. Let the ammeter be joined on to the + side, and suppose it shows a current of ·8 of an ampere. This means an earth on the − side, shown by dotted lines between P_1 and earth, which thus allows a small current to flow from the + to the − side ; and since it is assumed that the resistance to this leakage current is between earth and the − side, hence it measures the insulation resistance on the − side ; the resistance of the ammeter being extremely small, it may be neglected. Applying Ohm's law, this insulation resistance is obtained by dividing the pressure of the circuit by the leakage current. Suppose the circuit pressure is 110 volts, the leakage to earth on the − side is ·8 of an ampere, then the insulation resistance of the − side will be $110 \div ·8 = 137$ ohms. For this test very low-reading ammeters are required, ranging from 0 to, say, six or seven amperes, with readings of one-tenth of an ampere, and these are not met with in central stations. The following test, however, is one that can always be made. It consists in simply supplanting the ammeter by a voltmeter, the only thing that is necessary to be known being the resistance of the voltmeter.

Diagram III. illustrates the connections. If there is an earth on the − side there is a deflection of the voltmeter when joined up to the + side, the result of a small current flowing from the + side through the voltmeter, and then from earth to the − side. The total resistance between + and − side consists of the resistance of the voltmeter plus the resistance between earth and the − side, this latter forming the insulation resistance of the − side.

Let r = voltmeter resistance ;
  R_i = insulation resistance ;
  E = pressure of the circuit in volts ;
  D_1 = deflection of voltmeter across mains ;
  D_2 = deflection of voltmeter across mains when earthed.

The deflection obtained will be inversely proportional to the total resistance.

Hence $$\frac{D_1}{D_2} = \frac{R_i + r}{r};$$

but D_1 = E because it indicates numerically the voltage of the circuit. Therefore, supplanting D_1 by E, we have

$$\frac{E}{D_2} = \frac{R_i + r}{r}; \quad \therefore R_i = \frac{E\,r}{D_2} - r.$$

Taking a practical case, let the pressure of the circuit be, say, 110 volts, the resistance of the voltmeter 2,000 ohms, and suppose we obtain a deflection of 12deg. when connected between the + omnibus bar and earth, then

$$R_i = \frac{110 + 2,000}{12} - 2,000 = 16,333 \text{ ohms} ;$$

so that in this case the insulation resistance of the whole of the − side of the circuit comes out to 16,333 ohms.

Polarity Test.—Direct-current dynamos are liable at any time to have their polarity reversed. This is not often so ; still, there is the possibility of this happening, and it is well to have some reliable means of quickly detecting any change, should such change occur. An overhead arc circuit, with wires exposed to the atmosphere, affords a ready medium for allowing a change of polarity to take place in the dynamo connected to that circuit, and often brought about by the action of lightning. In the case of an arc circuit, the result would be that the arc lamps would burn upside down, while on an incandescent circuit it would not signify. But it is when a group of dynamos are connected up so as to run in parallel that the polarity of the field becomes a matter of serious importance, because if a dynamo whose polarity is reversed is put into circuit with another it will be in series instead of in parallel, consequently there will be double the pressure on the lamp circuit.

Fig. 47 illustrates a very simple and thoroughly practical way of determining whether the polarity of any dynamo is reversed or not. This device is due to Mr. L. Le Gros Horne, and the diagram almost explains itself.

On a parallel switchboard it is usual to have a starting voltmeter, which can be connected up to any individual machine. The device in question simply consists in placing an incandescent lamp between each omnibus bar and each terminal of the voltmeter, so that one lamp is placed between the positive omnibus bar and the positive terminal, and

the other lamp between the negative omnibus bar and the
negative terminal of the voltmeter.   When the starting
voltmeter is not connected up, then a feeble current will flow
through the lamps, since the two lamps and the voltmeter
are all in series.   By connecting up the voltmeter to the
machine that is to be put into parallel, the lamps will glow
at half voltage each as the current flows through the two
lamps (and through the armature of the machine, if the

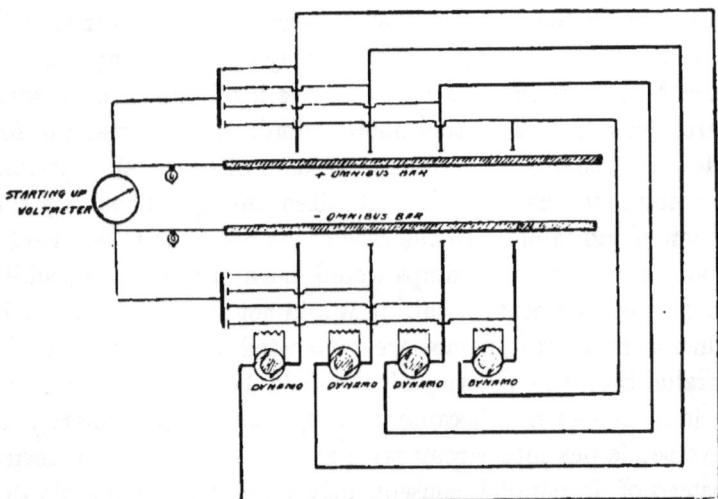

FIG. 47.

brushes of the latter have been put down), then as the dynamo
excites up so the light in the lamps will gradually die out, until,
when the machine is at full voltage, there will be no current in
the lamps, and consequently no light, because the pressure
of the dynamo will oppose the pressure of the circuit.   In
the event, however, of the polarity of the machine being
opposite to the polarity of the circuit, then the pressure of
the machine will be added to the pressure of the circuit, and
thus produce double the working pressure ; and since the two

lamps are in series this will bring them both up to full candle-power, as two lamps in series require double the pressure of a single lamp. So that the glowing of the two lamps will produce a visible danger signal and warn the attendant that the dynamo has its polarity reversed ; if the lamps show no light, then all is right. It is necessary to watch when the voltmeter is first connected up, and before the dynamo has begun to excite, and observe whether the lamps give a red glow at half voltage, because if not, then one lamp will probably be broken. If this is not discovered, the absence of light later on would be mistaken for showing right polarity, and in the event of the polarity of the machine having become reversed, the warning could not be given, owing to the broken lamp interrupting the circuit.

Traction Notes.

On first thoughts it may seem strange that, whilst a tramcar is ordinarily run by two horses, a third being generally attached for going up steep gradients, yet an electric tram line has two electromotors in each car, each perhaps capable of developing $12\frac{1}{2}$ b.h.p , or a total of 25 b.h.p. when working at its maximum efficiency. The explanation of this may be traced to two chief facts : first, the great variation of efficiency which electromotors possess for various speeds ; second, the undue amount of initial pull or torque required to start the car from rest and to get it up to speed. A horse by exertion can put forth double or treble its normal tractive power for a few seconds, just long enough to start the car and get it under way, as anyone can observe ; but when dealing with machines the case is different : they will only give out the maximum power for which they are built—that is, taking their utmost maximum—and with a series motor its useful power is limited by its temperature, and if more than the safe maximum power be drawn from it, it will simply

burn up.   Hence, in the first place, a motor must exert a
very strong torque in order to start the car, and to do this
its armature must be built so that it can carry a heavy
current for a length of time necessary to get up the speed of
the car, perhaps six or eight miles per hour.   At the instant of
starting, and for three or four seconds afterwards, the motor
may have to receive a current ten times the amount it would
receive when the car is running at its normal speed on a level
track, but although a great amount of power is thus *absorbed*
by the motor, yet it *yields* only a small power, because the
speed is so slow, the car being just on the move; hence the
efficiency of the motor when thus working may be only from
5 to 10 per cent.

In order to calculate out the power requisite for running
any vehicle, it is necessary to know the coefficients of friction
for various conditions of the road.   These coefficients of
friction are usually worked out in pounds per ton weight,
and for a carriage running on rails about 20lb. per ton may
be taken as a fair value.   It is very easy to obtain the horse-
power necessary to propel a car.   Given the weight of the
car and the speed ; each ton weight of the car signifies 20lb.
of resistance, and expressing the speed as feet per minute,
then the product of the total resistance and the speed per
minute in feet will give the foot-pounds of work requisite to
draw the car along at that speed ; hence dividing by 33,000
will give the horse-power.   For example, suppose we require
to find the horse-power necessary to draw a car at eight miles
per hour, the weight of the car being 12 tons, and the friction
on the rails being 20lb. per ton.   When it is stated that the
frictional resistance of the car is 20lb. per ton weight, this
signifies that pulling or pushing the ton weight on level rails
is equivalent to lifting vertically against gravity a weight of
20lb. ; hence the total resisting force offered by the car will
be $12 \times 20 = 240$lb.   Eight miles per hour $= 704$ft. per

minute, and to exert a pull or force of 240lb. through a distance of 704ft. requires 240 × 704 = 168,960 foot-pounds of work ; and since this is done in the space of one minute, hence the power exerted will be 168,960 ÷ 33,000 = 5·12 h.p.   This is only for level roads ; when there is a gradient to mount, an additional calculation must be made. Suppose that in the above case the car had to travel up an incline of 3 in 100—*i.e.*, every 100ft. produced a rise in eleva-tion of 3ft. — then in 704ft. the rise of level would be 704 × 3 ÷ 100 = 21·12ft.   Hence, the car would rise bodily a vertical distance of 21·12ft. when it has travelled 704ft.   The weight of the car is 12 tons = 12 × 2,240 = 26,880lb., and 26,880lb. raised 21·12ft. signifies 26,880 × 21·12 = 567,705 foot-pounds of work.   This is for one minute, and hence it requires 567,705 ÷ 33,000 = 17 h.p. The power required to run the car along a level is, say, 5 h.p., hence the total power required to run it up an incline of 3 in 100 will be 5 + 17 = 22 h.p.

At the central power station of an electric tram line the power required varies in a most remarkable degree, due to the sudden and irregular load that is thrown on the machinery.   This is greatly intensified when the line has a number of steep gradients.   At one instant the ammeters may show no current whatever, and the next hundreds of amperes.   This extraordinary variation is due to the sudden stopping and starting of cars on the road, and when several cars happen to start at the same moment, then the effect is still more marked.   All this variation of load signifies a great strain on the machinery in the generating station, because the engine governors have no time scarcely to act, and at one instant the engines may be working against a very heavy load, and in a second or two afterwards the load may be almost at zero.   It is for this reason that nearly all engines in power-houses drive the dynamos by belting, as the belt forms a somewhat elastic connection between the engines and dynamos, and so prevents the great shocks to either that would inevitably occur if there was a rigid connection, as direct coupling.

At the City and South London Electric Railway there are four dynamos, of the Edison-Hopkinson type, each having an output of 450 amperes at 500 volts pressure, and weighing 17 tons. They are driven by link leather belting 28in. wide, a jockey pulley being mounted close by the dynamo to assist the drive. The engines are of vertical compound type, giving 375 i.h.p. each, each engine driving its own dynamo, and running at a speed of 100 revolutions per minute. At the Liverpool Overhead Railway the driving is by ropes, and the engines are horizontal instead of vertical. On both lines the working pressure is 500 volts, constant-pressure system. It is, however, impossible to keep the pressure constant, owing to the enormous fluctuations of current. This pressure is found to be the most suitable for tractive purposes, since it is not too high to be particularly dangerous, and it is sufficiently high to obviate the use of very heavy currents on the line. Nearly all the American tram lines on the overhead trolley system work at 500 volts pressure.

There are, broadly speaking, three systems of electric traction : (1) the conduit system; (2) the storage battery system; (3) the overhead or trolley wire system. So far, no really successful method on the conduit system has been obtained, although numerous attempts have been made to solve the problem. By the " conduit system " is signified the supply of electric energy to the cars by means of an insulated cable positioned in the ground along the track, and making contact with the car motor as it travels along by means of a trailing brush or sliding contact. The supply cable is usually fixed in a conduit or channel, which is flush with the track, a slot being made in the conduit through which the brush makes contact with the supply cable. The fatal objection to this method of making contact is that the slot allows rain and dirt to enter, and so the slot is liable to be filled with water and mud, and thus the insulation of the supply cable is partially, if not wholly, destroyed. The general opinion is that an open conduit, as the slotted conduit is named, can never become a thoroughly reliable method of supplying electric power to electric cars. Numerous attempts have therefore been made to produce a closed conduit, thus obviating the disadvantages of the open conduit, but so far no system has been brought forward that has not some drawbacks to it ; and the advent of a perfectly reliable closed-conduit system would go a long way towards promoting the adoption of electric traction on our street tramways.

With regard to propelling cars by the use of storage batteries placed on the car, not very much success has been attained in this direction. The great weight of the batteries as compared with their output, the deterioration of the plates and the rough usage to which they must necessarily be subjected are all great drawbacks. Nothing damages a storage cell so much as to discharge it at a very high rate, and this is precisely what would be required if used to drive electric cars. The sudden starting of the cars, particularly if on a gradient, would in time ruin the cells. Improvements in storage cells are still going on, but they will require to be very much improved before they can be used as a cheap and satisfactory medium for tractive purposes.

The third system—namely, the overhead trolley wire system—has much to commend it. It is in use almost everywhere in the States, and has been adopted in several places in this country : the first being the Roundhay Electric Tramway at Leeds, then on the South Staffordshire tram lines.

The only objection that can be made against it is the appearance of a network of wires which span either across the road or half-way across, with the necessary supporting poles. It is possible, however, to make them look very trim, and the wires being thin do not present such a very objectionable appearance, after all. On the other hand, this system of working the cars has been brought to a high state of perfection, and the ease and reliability with which they run more than compensate for any imaginary æsthetic objections.

But perhaps the greatest field for electric traction at present is in the construction of tubular underground electric railways similar to the City and South London Electric Railway. The traffic of a city like London is already so enormous that the streets are almost impassable at times, and the construction of tubular electric railways underneath our main thoroughfares would provide an excellent means of rapid passenger conveyance, and so tend to relieve the congested traffic in the streets.

Cost of Electric Energy.

Within the last few years there has been a considerable decrease in the cost of electric energy. It was not an uncommon thing for a public supply company to charge 1s. per kilowatt-hour, and now the ruling charge is only 6d.; and as time goes on we may certainly look forward to a further reduction. Comparatively speaking, all electric supply companies are so far only on the threshold of their undertakings, and the largest of them does not supply more than 100,000 lamps or so, and this represents a small quantity of light compared to what is supplied by a gas company. The greater the scale upon which electricity is generated the cheaper does it become, and with this fact in view, it is not saying too much to state that there is great probability that we shall in time be able to procure electric energy at a lower rate than at present. Large private establishments can generate their own electricity for a very small sum. At present supply companies are burdened by the heavy expense incurred ,in laying down miles of distributing mains over wide areas, and in return they only supply a small percentage of houses with light ; but as the electric light gradually supplants gas, then these mains will become worked at higher efficiency. Another serious obstacle to the low price of electricity is the absence of a day load. For a few hours in the evening the machines are worked at their full load and efficiency ; after midnight only a very small load is on ; and if winter time, a fair load is obtained in the early morning—the rest of the day is almost unproductive. A foggy day is a source of great revenue to an electric supply company, since the machinery is kept at work all day and evening. It is hoped to remedy this deficiency of load in the daytime by supplying current to electromotors situated in workshops, factories, etc. ; then when the day comes to a close and work stops, the power that supplied current to the machines can then supply current for lighting purposes, but at present the supply of current for motive power is extremely small, electromotors not yet having come into general use.

In large private premises, such as mills, business establishments, etc., the use of the electric light has been found to be cheaper than gas—in some cases 50 per cent. as cheap. Where a large amount of steam power is in use, the comparatively small amount taken by a dynamo does not entail much expense, because the steam boilers and engines are used for driving the mill machinery, and 1,000 h.p. is considered nothing out of the ordinary in mill work. Now, 10 per cent. of this power—namely, 100 h.p.—will run, say, 900 lamps of 16 c.p. each—a quantity sufficient to light a fair-sized place. The extra amount of coal and water consumed to produce the necessary amount of electrical energy is only a moderate amount when judged with the amount required to drive the mill machinery.

The attendance required need only be small, and that is limited to the electrical part of the plant, since the care of the boilers and engines is left to the usual mill staff. To put down a special plant in any premises, consisting of motive power in addition to the electrical equipment, is very expensive in first cost or outlay, and it also costs a fair amount to maintain in proper working order, since a stoker and driver, in addition to the dynamo attendant, is required ; but even with this heavy expense, in certain large works and establishments it is found cheaper than using gas.

The following details of a gas-engine plant may be interesting, as showing the difference between burning gas direct to produce light and burning it in a gas-engine to produce light by means of electric energy. The data given embody average results extending over a period of six months' run, the figures having been taken by the author while running the plant. The dynamo had an output of 90 amperes at 100 volts, and so was a nine-unit machine ; it was driven by one of Crossley's new high-speed gas-engines, F size, guaranteed to work to 14 b.h.p., the engine being fitted with two heavy flywheels, 5ft. 2½in. diameter with face of 8½in. These wheels were weighted near the boss with a large mass of metal, and their large size and high speed gave a great steadiness to the running of the engine. The dynamo was driven by a long heavy link leather belt, the distance between the centres of the engine driving wheel and the dynamo pulley being about 20ft. Owing to the length of drive and weight of belt, the impulses due to the explosions were not transmitted, and the running of the dynamo was as steady as if it were driven by a steam-engine. In a

great number of cases, each explosion in the cylinder of a gas-engine produces a momentary flicker or pulsation of the light in the lamps, which is highly objectionable, and by this it is very often easy to tell that the light is produced by a gas-engine plant. In the above case the three things that cured these defects were :

1. The regular running of the engine.
2. The high speed and weight of the two flywheels.
3. The long drive by a heavy belt.

The stored energy in the flywheels prevented the periodic effect of the impulses from being felt. The plant ran for about 10 hours per day, except Sundays. The dynamo having a constant load of 9,000 watts, allowing a loss of 10 per cent. between the engine driving wheel and the dynamo terminals, part of this being due to the belt, the author calculated that fully 14 b.h.p. was developed by the engine. With 60 watts per incandescent lamp, the regular lamp load was about 150 lamps of 16 c.p.; this gives $150 \div 14 = 10{\cdot}7$, or over 10 lamps per brake horse-power of the gas-engine—an excellent performance.

Towards the evening it was found that the load could be slightly increased by adding more lamps, on account of the pressure of the gas being better. The consumption of gas was, over an average of six months, at the rate of 90,000 cubic feet per month, and price was 2s. 9d. per 1,000 cubic feet, so that the cost per month was £12 7s. 6d., or about £150 per annum. This gives the cost of gas alone for one incandescent lamp of 16 c.p. per annum as $150 \div 150 = £1$; 14 b.h.p. for 3,000 hours signifies 42,000 horse-power hours ; 90,000 cubic feet per month =, say, 1,080,000 cubic feet per annum, and 1,080,000 divided by 42,000 gives us 25·75, or $25\frac{3}{4}$ cubic feet of gas consumed by the gas-engine per brake horse-power. A 14-c.p. gas jet takes about five cubic

Y

feet per hour, therefore 25¾ cubic feet would yield a fraction over five gas jets, or a total illuminating power of 5×14 = 70 c.p., but this same 25¾ cubic feet of gas, if put into a gas-engine and burnt, and allowed to do work by driving a dynamo, will give 1 b.h.p., or sufficient electrical energy to run 10½ incandescent lamps of 16 c.p. each, or a total illuminating power of $10\frac{1}{2} \times 16 = 168$ c.p., so that when burnt in a gas-engine a certain quantity of gas will yield more than twice as much (or as 168 : 70) illuminating power than if it were burnt direct in a gas jet. Putting this in another way, gas burnt from a gas jet only gives 41 per cent. of the light that would be given when burnt in a gas-engine to produce electricity.

INDEX.

♦♦♦♦♦♦♦

Y 2

PRINTED BY BIGGS AND CO., SALISBURY COURT, FLEET STREET, E.C.

PRACTICAL ELECTRICAL ENGINEERING: Being a complete treatise on the Construction and Management of Electrical Apparatus as used in Electric Lighting and the Electric Transmission of Power. 2 Vols. Imp. Quarto.

By VARIOUS AUTHORS.

With many Hundreds of Illustrations. PRICE £2 2s. 0d.

Among the information collected in and written specially for these volumes, are complete monographs by

MR. GISBERT KAPP, M.Inst.C.E., on Dynamos, giving the principles on which the construction of these machines is founded, and a vast amount of new information as to the constructive details.

MR. ANTHONY RECKENZAUN, M.Inst.E.E., considers the whole question of Electric Traction on Railways and Tramways, describing the most recent practice and giving exhaustive details not only of the mechanism, but of practical considerations which are of the utmost commercial importance.

MR. C. CAPITO, M.Inst.E.E., deals with Steam—and considers the theory—as well as the practical construction of boilers and engines, more especially in connection with electric lighting and transmission of power. This section alone contains no less than 371 illustrations.

MR. HAMILTON KILGOUR considers carefully the theory of Electrical Distribution, upon which depends the right proportioning of mains in order to secure economical distribution. Other men have collected a complete description of the various systems practically applied, the materials used and methods of construction.

The above form but a portion of these volumes, but should prove of the utmost value to everyone who professes to be an Electrical Engineer.

BIGGS & CO., 139-140, SALISBURY COURT, LONDON, E.C.

DYNAMOS, ALTERNATORS, AND TRANSFORMERS.

By GISBERT KAPP, M.Inst.C.E., M.Inst.E.E. Fully Illustrated. Price 10s. 6d.

The book gives an exposition of the general principles underlying the construction of Dynamo-Electric Apparatus without the use of high mathematics and complicated methods of investigation, thus enabling the average engineering student and the average electrical engineer, even' without previous knowledge, to easily follow the subject.

We can heartily recommend it.—*Electrical Engineer.*
A valuable contribution to electrical literature.—*Electrical World.*
Invaluable to the advanced student and dynamo designer.—*Electrician.*
The reader will find valuable information concerning dynamo design.—*Nature.*

BIGGS & CO., 139-140, SALISBURY COURT, LONDON, E.C.

ELECTRIC TRACTION. By A. Reckenzaun, M.I.E.E.

Illustrated. Price 10s. 6d.

This is the first serious attempt to consolidate and systematise the information of an important subject. Mr. Reckenzaun's experience is of the longest and widest, and this book deals not only with the scientific and practical problems met with in traction work, but enters somewhat into the financial aspect of the question.

CHAPTER I.—Early History. Magnetic Fields. Torque. Motor Efficiency. Insulation Resistance. Brake Tests. Electrical Horse-Power.

CHAPTER II.—Traction. High Pressure. Current, Stopping and Starting. Current Running. Energy Used.

CHAPTER III.—Advantages of Two Motors on Each Car. Calculations for Armatures. Constructive Details for Motors. Types of Tramcar Motors. Method of Suspension. Brushes and Brush-holders. Switches and Speed-Regulating Devices. Motor Trucks. Important Points in Tramcar Motors. Gearless Motors. Mechanical Transmission between Motor and Axle.

CHAPTER IV.—Description of Principal Systems of Overhead and Underground Construction used in Electric Traction. Poles. Trolleys and Trolley Wire. Overhead Curve Construction. Practical Hints on Overhead Construction. Insulators used in Electric Railway Work. Switchboards. Lightning Arresters and Cut-Outs.

CHAPTER V.—Secondary Batteries. Weight and Efficiency Deterioration.

CHAPTER VI. Portrush, Bessbrook-Newry, City and South London Railways. Central London Railway, Blackpool, Budapest, Leeds, and Halle Electric Tramways, West End Railway, Boston.

CHAPTER VII.—Résumé of Writings and Expressions of Prominent Electrical Engineers on Electric Traction.

CHAPTER VIII.—Details of Working Expenses.

Will be of great use as a reference to all tramway engineers, and will be invaluable to electrical engineers commencing traction work.—*Electrical Engineer*.

His book is certainly interesting and instructive.—*Electrical Review*.

The most useful to English readers.—*Engineer*.

BIGGS & CO., 139-140, SALISBURY COURT, LONDON, E.C

ELECTRICAL DISTRIBUTION: ITS THEORY AND PRACTICE. Part I.: By Martin Hamilton Kilgour. Part II.: By H. Swan and C. H. W. Biggs. Illustrated. Price 10s. 6d.

PART I.

CHAPTER I.—Introduction.

CHAPTER II.—Constant-Current (Series) System. Constant-Current (Parallel) System. Feeders. Distributing Mains. Supply to Apparatus at a Variable Distance from the Generating Source.

CHAPTER III.—Economy in Design. Supply to Apparatus at a Constant Distance from the Generating Source.

CHAPTER IV.—Miscellaneous Problems on Feeders. Distributing Mains.

CHAPTER V.—Hypothesis. Calculations. Tables.

CHAPTER VI.—Examples. Tables of Square Roots. Particulars of Conductors.

PART II.

CHAPTER VII.—Bare Wire and Modification of Bare Wire. Crompton's System. Kennedy's System. St. James's and Pall Mall. Tomlinson's System.

CHAPTER VIII.—Practice in Paris. Compagnie d'AirComprimé et d'Electricité. Continental Edison. Société du Secteur de la Place Clichy. Secteur des Champs Elysées. Halles Centrales.

CHAPTER IX.—Callender-Webber System. Callender Solid System.

CHAPTER X.—Brooks Oil Insulation System.

CHAPTER XI.—Ferranti Concentric Mains.

CHAPTER XII.—Modified Systems—Silvertown. St. Pancras Mains.

CHAPTER XIII.—Johnstone's Conduit System.

CHAPTER XIV.—Crompton and Chamen with Accumulators.

Mr. Kilgour's treatment of his subjects will commend itself to all who are interested in them.—*Engineer.*

An excellent compendium on the subject.—*Electrical Engineer.*

Of high interest and usefulness.—*Nature.*

BIGGS & CO., 139-140, SALISBURY COURT, LONDON, E.C.

ELECTRIC LIGHT AND POWER: Giving the Results of Practical Experience in Central-Station Work. By ARTHUR F. GUY, A.M.I.C.E. Illustrated. Price 5s.

This book is issued for the purpose of placing on record useful practical knowledge obtained by the author during several years' experience of central-station work, together with brief explanations of the laws which govern the action of electrical apparatus in general use for electric lighting. The contents comprise:

CHAPTER I.—Evolution of Electrical Engineering: The First Electric Light—Subsequent Progress—Electricity in England and Abroad—Economics of the Electric Light—Sources of Power—Conservation of Energy, etc.

CHAPTER II.—Motive Power: Coal as Fuel—Plotting Curves— Steam Boilers, Engines, etc.—Running Cost of Steam and Gas Engines—Dowson and other Gas Producers— Petroleum Oil-Engine—Water Power, etc.

CHAPTER III.—Practical Laws of Electricity and Magnetism: Sources of Electricity—Electrical Units—Calculation of Resistance—Conductors and Insulators—Simple and Divided Circuits — Magnets and Magnetism — The Magnetic Circuit, etc.

CHAPTER IV.—Electric Machinery: Generation of Current— Field Magnet and Armature Winding—Working in Parallel—Notes on Running—Cost and Output of Dynamos—Electromotors, etc.

CHAPTER V.—Electric Arc and Incandescent Lighting: Illuminating Power of Arcs—Consumption of Carbon— Diffusion of Light—Fixing and Trimming—Arcs in Parallel — Town-Lighting — Life and Efficiency of Lamps, etc.

CHAPTER VI.—Distribution of Electric Power: Low-Pressure System—Loss in Mains—High-Pressure System— Use of Three and Five Wires—Alternate-Current Working—Transformers—Insulation—Testing Circuits— Traction Notes—Cost of Electric Energy, etc.

BIGGS & CO., 139-140, SALISBURY COURT, LONDON, E.C

ECONOMICS OF IRON AND STEEL. By H. J. Skelton. Illustrated. Price 5s.

BIGGS & CO., 139-140, SALISBURY COURT, LONDON, E.C.

FIRST PRINCIPLES OF MECHANICAL ENGINEER-
ING. By JOHN IMRAY, with additions by C. H. W. BIGGS. Illustrated. Price 3s. 6d.

An attempt is made in this book to explain the principles of Mechanical Engineering in simple language, and without the aid of abstruse mathematics. The calculations are generally arithmetical, and such as come well within the comprehension of the beginner.

CHAPTER I.—Mechanics of Antiquity. Modern Machinery. Mechanics and Chemistry. Statical and Dynamical Machines. Strength of Materials. Nomenclature. Unit of Work. Source of Energy. Application of Energy. Friction. Governing. Nature of Machines. Mechanical Education.

CHAPTER II.—Mechanical Drawing. Use of Drawing. Plane Surfaces. Drawing Cubes. Drawing Cylinders. Projection. Sections. Perspective, etc. Instruments. Scale of Drawings.

CHAPTER III.—Strength of Materials. Strains and Stresses. Tension. Compression. Transverse Stress. Deflection and Disposal of Materials. Calculations. Torsion. Shafts. Clipping and Shearing Stress.

CHAPTER IV.—Sources of Mechanical Power. Muscular Force. Wind. Windmills. Water. Waterwheels. Turbines. Weight. Falling Bodies. Springs.

CHAPTER V.—Heat. Expansive Power of Heat. Elasticity of Gases. Temperature and Pressure. Condensation. Expansion. Steam Boilers. Steam Engines.

CHAPTER VI.—Electricity. Chemical Action.

CHAPTER VII.—Transmission of Power. Rotary Motion. Couplings. Clutches. Plummer Blocks. Pulleys or Drums. Toothed Wheels. Reciprocating Motion. Discontinuous Motion. Ratchet. Cam. Mangle Motion. Reversing Gear. Dynamometer. Brakes, etc.

CHAPTER VIII.—Transmission of Power. Water. Compressed Air. Electricity.

The book will be found useful to learners.—*Engineer.*

The poor student of to-day is fortunate to have at his command this useful volume at so low a price.—*Contract Journal.*

Written in a clear and simple manner.—*Builder.*

BIGGS & CO., 139-140, SALISBURY COURT, LONDON, E.C.

FIRST PRINCIPLES OF BUILDING: Being a Practical Handbook for Technical Students.

By ALEX. BLACK, C.E. Illustrated. Price 3s. 6d.

The following table of contents will show the matters treated in this work, and it will, we think, be found that a large amount of information is given herein that is not usually found except in expensive treatises :

BIGGS & CO., 139-140, SALISBURY COURT, LONDON, E.C.

FIRST PRINCIPLES OF ELECTRICAL ENGINEER-
ING. Second Edition. Crown 8vo. Price 2s. 6d.
By C. H. W. Biggs, Editor of *The Electrical Engineer* and *The Contract Journal.*

The first edition of this book has long been out of print, but the author has been unable owing to pressure of other work to arrange for the publication of the new edition. The original book was, as was expected, severely handled by the critics. As a matter of fact it was intended to be a bone of contention, and the author has found no reason to withdraw from the position he then took up. Some portions of the book have been rearranged, and in one or two cases where condensation led to the views promulgated being mistaken, the matter has been expanded and more fully illustrated. The book is intended to be a first book for electrical engineers, and avoids as much as possible the discussion of electrical questions that have no bearing upon the probable future work of the reader.

CHAPTER I.—Introductory.

CHAPTER II.—Deals with the Conductive Circuit, the Inductive Circuit, and the Magnetic Circuit.

CHAPTER III.—Discusses the Production of Electrical Pressure or Difference of Pressure. The Use of Electro-Graphics, Loops of Force, and Interaction among Circuits.

CHAPTER IV.—Kinds of Dynamos. Characteristic Curves. Self-induction. Motors. Distribution. Simple Measurements. How to Measure Current. The Tangent Galvanometer. Ampere-meters and Voltmeters. Cardew Voltmeter. Wheatstone Bridge Method.

The first principles of the dynamo are clearly and accurately given.—*Nature.*

We commend the book to the perusal of students.—*Electricity.*

BIGGS & CO., 139-140, SALISBURY COURT, LONDON, E.C.

PORTATIVE ELECTRICITY. By J. T. NIBLETT, author of " Secondary Batteries." Illustrated. Price 2s. 6d.

This is a treatise on the Application, Method of Construction, and the Management of Portable Secondary Batteries.

INTRODUCTION.—A brief *résumé* of Electrical Discovery.

PART I.—Portative Electricity in Mining Operations, describing some of the Lamps used, and Instruments for the Detection of Dangerous Gases. Safety Electric Hand-lamps for Domestic Purposes, for Customs' Officers, for Meter Inspectors, and for Firemen. Portative Electricity for Domestic Purposes, for Reading, for Driving Motors, for Sterilising Water. Application to Medical and Scientific Purposes. Uses for Military Purposes, for Land Work, At Sea. Uses for Lighting Vehicles, Railway Carriages, Omnibuses, Tramcars, Broughams, Cycles. Traction by means of Portative Electricity. Applications to Decorative Purposes, for Dinner Tables, Shop Windows, Fountains, Jewellery, on the Stage, Personal Adornment.

PART II.—Secondary Cells, Planté's, Crompton-Howell, Epstein, D. P., Faure, E. P. S., Lithanode, Pitkin, Bristöl. Types of Secondary Cell. Niblett's Solid Cell.

PART III.—The Management of Portable Apparatus for storing Electrical Energy. Methods of Developing Electrical Energy. Thermopiles. Primary Batteries. Charging Instructions.

The Charging Instructions are very complete, and should be of immense service to those who use small Portable Cells.

Will be more especially valued for its practical facts and instructions.—*Electrical Engineer.*

Contains much useful and interesting matter.—*Glasgow Herald.*

The book is written in a popular style.—*Mechanical World.*

May be regarded as a veritable premonition of the dawn of the " electric age."—*Liverpool Courier.*

BIGGS & CO., 139-140, SALISBURY COURT, LONDON, E.C.

POPULAR ELECTRIC LIGHTING. By Captain E. Ironside Bax. Illustrated. Price 2s.

It is well written, in a popular style, and explains very clearly the general points in connection with electric lighting.—*Electrical Review.*

BIGGS & CO., 139-140, SALISBURY COURT, LONDON, E.C.

ALTERNATE-CURRENT TRANSFORMER DESIGN.

By R. W. WEEKES, Whit.Sch., A.M.I.C.E. Crown 8vo.
Price 2s.

This book is one of a new series intended to show engineers and manufacturers the exact method of using our acquired knowledge in the design and construction of apparatus. Mr. Weekes has taken a number of different types and calculated out fully the dimensions of the various parts, showing each step in the calculation. Diagrams are given drawn to scale, and a summary of sizes, weights, losses, and costs given at the end of each design.

At the present time transformers play an important part in high pressure distribution, and it is of the greatest importance that they be constructed to give as little loss as possible.

We commend it to the notice of our readers.—*Electricity*.
This little work will much assist in imparting a knowledge of the designs treated of.—*Electrical Review*.

TOWN COUNCILLORS' HANDBOOK TO ELECTRIC LIGHTING. By N. SCOTT RUSSELL, M.Inst.C.E.
Illustrated. Crown 8vo. Cloth, 1s.

This work is intended to afford to County Councillors and others who are interested in electricity some information likely to be of use to them in dealing with questions of central-station lighting, and to convey that information in as simple and popular a form as possible.

The great interest taken in electric lighting by municipal and other local authorities leads, not only the members of these bodies and the officials, but a good many of their constituents, to desire some elementary information on the subject. This, Mr. Scott Russell provides, and tells how to go to work to obtain provisional orders, as well as simply discusses the value of this system of lighting.

Seems to have accomplished the object in view.—*Nature*.
Has done yeoman service in preparing this little book.— *Electrical Engineer*.
A useful shilling handbook that every town councillor should read.—*Building News*.

BIGGS & CO., 139-140, SALISBURY COURT, LONDON, E.C.

THEORY AND PRACTICE OF ELECTRO-DEPOSITION : Including every known Mode of Depositing Metals, Preparing Metals for Immersion, Taking Moulds, and Rendering them Conducting. By Dr. G. GORE, F.R.S. Illustrated. Crown 8vo. Price 1s. 6d.

This little work is too well known to need a lengthy description. It has long been acknowledged to be the simplest elementary manual on the subject, and Dr. Gore's ability and great care to put his matter clearly is nowhere better shown than in this little book.

PRACTICAL INSTRUMENTS FOR THE MEASUREMENT OF ELECTRICITY. By J. T. NIBLETT and J. T. EWEN, B.Sc.

The aim of the authors is to describe the theoretical considerations involved in the construction of the various electricity measuring instruments, the methods employed in their manufacture, and their commercial applications.

THE MARINE ENGINEER'S ELECTRICAL POCKET BOOK. By M. SUTHERLAND (Electrical Engineer at W. Denny and Bros.' Shipbuilding Yard, Dumbarton).

The writer has frequently been asked by marine engineers to recommend a book which would give them a sufficient insight into electrical engineering to enable them to understand ship installations without going too deeply into electrical matters in general, and though there are many excellent books on practical electrical engineering, he has not been able to find one which exactly fulfils these conditions. His object, therefore, has been to produce a compact and handy volume which should give an insight into magnetism and electricity, with descriptions of the various systems of wiring, generating plant, fittings, measuring instruments, etc., used on board ship, avoiding as much as possible all matter not directly applicable to work of this nature.

BIGGS & CO., 139-140, SALISBURY COURT, LONDON, E.C.

REFUSE DESTRUCTORS,

WITH

RESULTS UP TO PRESENT TIME.

SECOND AND REVISED EDITION.

A Handbook for Municipal Officers, Town Councillors, and others interested in Town Sanitation,

BY

CHARLES JONES, M.Inst.C.E.,

Hon. Secretary and Past-President of the Association of County, Municipal, and Sanitary Engineers and Surveyors; Surveyor to the Ealing Local Board.

WITH A PAPER ON

𝕮𝖍𝖊 𝖀𝖙𝖎𝖑𝖎𝖘𝖆𝖙𝖎𝖔𝖓 𝖔𝖋 𝕿𝖔𝖜𝖓 𝕽𝖊𝖋𝖚𝖘𝖊 𝖋𝖔𝖗 𝕻𝖔𝖜𝖊𝖗 𝕻𝖗𝖔𝖉𝖚𝖈𝖙𝖎𝖔𝖓,

BY

THOMAS TOMLINSON, B.J., A.M.I.C.E.

WITH NUMEROUS DIAGRAMS.   PRICE 5s.

MR. JONES is a well-known authority on all that concerns Refuse Destructors. He was among the first to suggest their use, and the addition of his own improvements made it possible to erect and use them in populated districts without creating a nuisance. In this book he has collected the information which his experience decides ought to be known, and has gone very fully into many details that are of the greatest importance to all officials and members of local authorities dealing with this question.

The book contains a large number of illustrations showing elevations and sections of many destructors.

Among the Contents will be found: Introduction, Fryer's Destructor, Healey's Destructor, Whiley's Destructor, the Horsfall Destructor, Boulnois and Brodie's Charging Apparatus, the Fume Cremator, Bradford Refuse Destructors, Leicester Destructors, Cost of Construction, Objections to Use, House Refuse and Sewage Sludge, Utilisation of Town Refuse, the Powdershall Destructor, Bradford Refuse Disposal, numerous Tabular Statistics, Disposal of House Refuse, Ealing, Liverpool Refuse Disposal, Tables of Answers on Refuse Disposal from various Towns, about twenty-five folded Plates of Illustrations of various Destructors, Statistics, etc.

BIGGS & CO., 139-140, SALISBURY COURT, LONDON, E.C.

MUNICIPAL ENGINEERS' SERIES.

——o——

TO BE PUBLISHED SHORTLY.

ROAD CONSTRUCTION AND MAINTENANCE. By H. P. BOULNOIS, M.Inst.C.E., Past-President of Municipal and County Engineers; City Engineer, Liverpool.

HOUSE DRAINAGE. By W. SPINKS, A.M.Inst.C.E., Lecturer on Sanitary Engineering, Yorkshire College, Victoria University.

DESIGN AND DISCHARGING CAPACITIES OF SEWERS, with Rules and Tables. By SANTO CRIMP, M.Inst.C.E., Author of "SEWAGE DISPOSAL."

WATER SUPPLY IN RURAL DISTRICTS. By R. GODFREY, A.M.Inst.C.E., Surveyor to the Rural Sanitary Authority, King's Norton.

HIGHWAY BRIDGES. By E. P. SILCOCK, A.M.Inst.C.E. Borough Surveyor, King's Lynn.

CONTRACTORS' PRICE-BOOK.—It would be difficult to prove that a want exists in any direction; but, as a matter of fact, while there are price-books in all directions and, seemingly, of every kind, it has been found from practical experience that not one of these books is of much assistance to large contractors—by large contractors, we refer to such as are engaged in railway, dock, sewage, canal, and other water works. This book is compiled to supply their needs, and will be published, corrected to date, at the beginning of each year.

BIGGS & CO., 139-140, SALISBURY COURT, LONDON, E.C.

www.ingramcontent.com/pod-product-compliance
Lightning Source LLC
Chambersburg PA
CBHW021400210326
41599CB00011B/950